An Introduction to S and S-PLUS

Phil Spector
University of California at Berkeley

 Duxbury Press
An Imprint of Wadsworth Publishing Company
Belmont, California

Editor: Stan Loll
Editorial Assistant: Claire Masson
Production Editor: Julie Davis
Copy Editor: Judith Abrahms
Permissions Editor: Peggy Meehan
Cover Design: Cloyce Wall
Printer: Malloy Lithographing, Inc.

IP ™

International Thomson Publishing
The trademark ITP is used under license.

Duxbury Press
An Imprint of Wadsworth Publishing Company
A division of Wadsworth, Inc.

Printed in the United States of America

6 7 8 9 10-ML-05 04

Library of Congress Cataloging-in-Publication Data

Spector, Phil
    An introduction to S and S-plus / by Phil Spector
      p.   cm.
    Includes Index
    ISBN 0-534-19866-X
    1. S (Computer program language) I. Title
QA76.73.S15S63 1994
519.5'0285'5133—dc20   93-37391

Contents

Preface   **xi**

1 Introduction   1

1.1   Invoking S   1

1.2   Online Help   2

1.3   Data   3

1.4   The Basics of S   4

1.5   A Sample S Session   5

1.6   Using S with a Command File   9

1.7   Interrupting S   10

1.8   Accessing Your Operating System from S   11

1.9   Initialization and Wrapup   11

1.10   Options   12

     Exercises   *13*

2 Data in S   15

2.1   Information About Objects   15

2.2   Reading Data into S   15

2.3   Generating Data in S   18

2.4   How S Finds Data   21

2.5   Organizing Data   24
    2.5.1   Organizing Data: Lists   24
    2.5.2   Organizing Data: Matrices   25
    2.5.3   Organizing Data: Data Frames   28

2.6   Time Series   31

2.7   Complex Numbers   32

2.8   Libraries   33

2.9   Advanced Topics   34
    2.9.1   Attributes   34
    2.9.2   Access to Data   36
    2.9.3   Direct Access to Data   38
    2.9.4   Advanced Features of the scan Function   40

    Exercises   42

3   Operators and Functions
43

3.1   Arithmetic Operators   43

3.2   Logical Operators   45

3.3   Special Operators   47

3.4   Operator Precedence   47

3.5   Introduction to Functions   49

3.6   Mathematical Functions   50
    3.6.1   General Mathematical Functions   50
    3.6.2   Trigonometric Functions   51
    3.6.3   Functions for Complex Numbers   51
    3.6.4   Functions for Rounding and Truncating   52
    3.6.5   Mathematical Functions for Matrices   52

3.7   Functions for Sorting   55

3.8   Functions for Data Manipulation   57

3.9   Functions for Simple Statistics   60

3.10 Functions for Probability Distributions   63
      3.10.1 Probability Density (d)   63
      3.10.2 Probabilities (p)   63
      3.10.3 Quantiles (q)   64
      3.10.4 Random Samples (r)   64

3.11 Functions for Categorical Variables   66

3.12 Functions for Character Manipulation   69

      Exercises   73

4 Introduction to Graphics in S   **75**

4.1 Overview   75

4.2 High-Level Plotting Commands   76

4.3 Arrangement of Plots on the Page or Screen   77

4.4 Interacting with Plots   83

4.5 Commands for Graphics Examples   85

      Exercises   87

5 Subsetting and Reshaping Data   **89**

5.1 Subscripting   89
      5.1.1 Accessing Vector Elements   89
      5.1.2 Accessing Matrix Elements   94
      5.1.3 Accessing List Elements   101
      5.1.4 Accessing Data Frame Elements   103

5.2 Combining Vectors, Matrices, Lists, and Data Frames   105

5.3 Functions for Subsetting or Reshaping Matrices   110

      Exercises   112

6 Functions in S   **113**

6.1 Getting Started with Functions   113

6.2 Modifying an Existing Function Using `fix`   115

6.3   Modifying an Existing Function Using Editor
        Commands   116

6.4   Errors in Writing Functions   116

6.5   Creating a New Function   117

6.6   Checking for Function Name Clashes   119

6.7   Returning Multiple Values   119

6.8   Errors and Warnings   120

6.9   Local Variables and Evaluation Frames   121

6.10 Cleaning Up   122

6.11 Advanced Topics   123
        6.11.1 Variable Numbers of Arguments   123
        6.11.2 Retrieving Names of Arguments   125
        6.11.3 Operators   125
        6.11.4 Assignment Functions   126
        6.11.5 Default Values   128

6.12 Dynamic Loading of Outside Routines   130
        6.12.1 A Simple Example   131
        6.12.2 Communicating with S Functions   136
        6.12.3 More Complex Issues   139

6.13 Static Loading   144

        Exercises   144

7   Programming in S   147

7.1   Mapping Functions to a Matrix   147

7.2   Mapping Functions to Vectors and Lists   152

7.3   Mapping Functions Based on Groups   153

7.4   Conditional Computations   156

7.5   Loops   159
        7.5.1   for Loops   159
        7.5.2   while Loops   161

7.5.3   Control Inside Loops: **next** and **break**   162

7.5.4   **repeat** Loops   163

7.6   Advanced Topics   164

7.6.1   Calling Functions with Lists of Arguments   164

7.6.2   Evaluating Text as Commands   165

7.6.3   Object-Oriented Programming   165

Exercises   168

8   Printing and Formatting   171

8.1   Sending Output to a File   171

8.2   Writing S Objects for Transport   172

8.3   Command History   173

8.4   Command Re-editing   174

8.5   Customized Printing   174

8.6   Formatting Numbers   175

8.7   Printing Tables   177

8.8   Accessing the Operating System   179

Exercises   181

9   Advanced Graphics   183

9.1   Overview   183

9.2   Graphics Parameters   183

9.3   Layout of Graphics   186

9.4   Low-Level Plotting Commands   189

9.5   Annotating Plots with Text   190

9.6   Using the Plotting Commands   191

9.6.1   Multiple Lines or Groups of Points on a Plot   191

9.6.2   Legends   194

9.6.3   Multiple Plots with Identical Axes   196

9.6.4   Special Plotting Characters   198

9.6.5   Logarithmic Axes   198
9.6.6   Custom Axes   199
9.6.7   Customizing Barplots and Histograms   202
9.6.8   Annotating a Perspective (3-D) Plot   204
9.6.9   Drawing Diagrams   207

9.7   Postscript Device Driver in S-PLUS   207

9.8   Postscript Device Driver in S   209

9.9   Multiple Graphics Devices in S-PLUS   210

9.10   More Complex Layouts   211

Exercises   213

10   Statistical Models   215

10.1   Data for Statistical Models   216

10.2   Expressing a Statistical Model   217

10.3   Common Arguments to the Modeling Functions   219
10.3.1   `formula` Argument to the Modeling Functions   219
10.3.2   `data` Argument to the Modeling Functions   220
10.3.3   `subset` Argument to the Modeling Functions   221
10.3.4   `weights` Argument to the Modeling Functions   221
10.3.5   `na.action` Argument to the Modeling Functions   222
10.3.6   `control` Argument to the Modeling Functions   222

10.4   Using the Statistical Modeling Objects   223

10.5   Linear Models and Regression (`lm`)   224
10.5.1   Example: Linear Regression   225

10.6   Analysis of Variance (`aov`)   228
10.6.1   Example: One-Way Analysis of Variance   229
10.6.2   Example: Two-Way Analysis of Variance   230

10.7   Generalized Linear Models (`glm`)   231
10.7.1   Families, Links, and Error Distributions   232
10.7.2   Example: Logistic Regression   232
10.7.3   Example: Log-Linear Model   235

10.8   Generalized Additive Models (gam)   239
    10.8.1 Example: General Additive Model   239

10.9   Local Regression Models (loess)   242
    10.9.1 Example: Local Regression Model   242

10.10   Tree-Based Models (tree)   244
    10.10.1 Example: Tree-Based Model for Classification   244

10.11   Nonlinear Regression (nls)   246
    10.11.1 Example: Nonlinear Regression   247

10.12   Simple Statistical Inference   250

10.13   Multivariate Statistical Procedures   253

10.14   Function Minimization (ms)   256
    10.14.1 Example: Minimization   256

    Exercises   257

11   Applied Statistical Models   259

11.1   Graphics and Statistical Models   259

11.2   Sequential and Partial Sums of Squares   263

11.3   Contrasts and Model Matrices   264

11.4   Regression Diagnostics   267

11.5   Predicted and Fitted Values   268

11.6   Updating and Comparing Models   270

11.7   Stepwise Model Selection   273

    Exercises   276

Index   278

Preface

Purpose

This book is written to serve as an introduction for anyone who wants to learn how to use S or S-PLUS: students, engineers, statisticians, data analysts, or others who deal with data they wish to display, manipulate, analyze, or graph. The S language is a powerful tool for performing any of these tasks, and I've tried to present ideas and methods that will help you enjoy S and get the most out of its power. Every concept introduced is accompanied by an example. Some readers will find these examples sufficient to show them what they need to do to solve their particular problems. If you have a specific problem in mind as you read this book, I certainly encourage you to look for examples that are similar to yours and to copy them as best you can. On the other hand, if you're trying to learn the language from scratch, I hope the examples will make the ideas presented in the text more concrete, so that you can combine and refine them in ways you will find useful.

As a first pass through the book, I'd recommend reading all of Chapters 1 and 2 and the first five sections of Chapter 3 quite thoroughly. You can skim over the remainder of Chapter 3, looking for those sections that describe functions of interest to you. The sections in Chapter 3 that describe the various functions available in S present many of the basic tools that can be combined to perform more complex tasks; most users don't need to know about every one of them. Next, Chapter 4 should be read in its entirety, to give you an overview of the graphics capabilities of S.

The topics in Chapter 5 are important for a real understanding of how S works, and of how to manipulate data in S. The subscripting operations provide elegant and efficient ways to extract and modify data, and can often replace tedious programs that would be required in other languages. At this point, most of the basic tools you'll need to use S as an interactive computing language will have been introduced. The basics of graphics in S, outlined in Chapter 4, should get you started producing simple graphs, and you may want to skim Chapter 9 to find out about some of the more advanced capabilities of graphics in S that you might find useful.

You can learn more about writing functions, programming, and printing as your needs dictate. On the other hand, if you're interested in really improving your S skills, Chapters 6 and 7 may be the most useful and interesting chapters in the book. One of the greatest strengths of S is that it provides the ability to extend the language by writing your own functions, and, over time, you'll probably create a collection of functions that you'll use repeatedly to solve a variety of problems. In Chapter 7, I've tried to emphasize those programming skills that are specifically useful in S, in addition to presenting the tools that can be used for more traditional programming. Over time, you should discover your own programming style, which is most successful for you, by combining a selection of these techniques.

Chapter 10 presents the techniques used to perform a wide range of statistical modeling procedures. The first four sections of this chapter will provide the basics you need to use these statistical techniques; each of the later sections in this chapter addresses a particular type of statistical model, and these can be skimmed over until you find a specific section that is relevant to your own work. Each modeling function is accompanied by at least one example, and Chapter 11 provides some examples of ways that you can use the results of a statistical modeling procedure to gain further insights into your data. Like all the examples in the book, these either use data sets distributed with S or data sets that can easily be created from the distributed data sets, so you shouldn't have to enter any data to recreate the examples on your own computer. I strongly encourage you to try the examples on your own computer, and if a question comes to mind, to try it out on some data—either the data from the examples in the book, or data of your own. Your skill in using S (or any computer language) can advance only if you actually try the things you read about in books such as this one.

Advanced Topics

Some of the chapters have sections titled "Advanced Topics." As the name implies, these sections can simply be skipped on a first reading unless you're really serious about learning all the details of the language. You may find that after a while you'll come back to these sections when a particular problem presents itself.

On-line Help

I can't stress too strongly the importance of using the online help facility outlined in Section 1.2. This book does not attempt to document every capability or feature of every function in S, so if you're in doubt about the way to use a particular function, check the online help. I hope this book will provide you with pointers to the right functions to use, as well as a solid introduction to get you started with them.

Examples and Their Use

Throughout the book, examples are displayed in `typewriter` font; statements preceded with the greater-than sign (>) represent S statements typed in response to the S or S-PLUS prompt. Continuation lines (beginning with a plus sign (+)) are parts of S statements that have been split only to make them easy to read on the pages of this book, and don't actually have to be split if you're typing the example into your own computer. So if an example is displayed as

```
> plot(Income, Illiteracy, main="Plot of Two Variables",
+       xlab="Income",ylab="Illiteracy")
```

you can just as effectively type it in all on one line, without including the plus sign.

Some examples produce a lot of output, which it would not be appropriate to include in this book. In these cases, an ellipsis (three periods) will appear in the example output, like this:

. . .

If you type in an example that has an ellipsis in its output, you should expect to see more output in addition to what is shown in the book.

For the Experienced User

Finally, if you're a more experienced S user, and are searching for the solution to a particular problem, the table of contents and index may point you to the right place. Most of the sections are designed to be readable even if you haven't read everything that precedes them; when this is not the case, cross-references are provided.

1 Introduction

The S programming language provides an interactive computing environment for programming, graphics, and statistical analysis. A variety of data can be stored, manipulated, plotted, and analyzed using both built-in functions and user extensions. S can easily accommodate and manipulate data stored in ways suitable for direct manipulation as vectors and matrices, as well as the "observations and variables" model used in most statistical packages. This flexibility allows you to perform a wide range of tasks from within S. In addition, the extensibility of S allows you to add your own functions to the language, either to customize existing capabilities or to enhance the language with new capabilities. These functions can be written in the S programming language or in C or FORTRAN, or they can simply be interfaces to existing programs in the UNIX environment. Thus, whether you are using S as a calculator, as a platform for testing new techniques, or as a statistical analysis and graphics tool, the functions you need to get the job done either will be readily available or should be easy to add to the system.

The first chapter of this book will introduce you to the various types of data that can be stored within S, and to the various ways of organizing and referring to your data. Later chapters will explore data manipulation, graphics, programming, and data analysis. At each step, a variety of examples will be presented. In order to really learn any computer language, the most important thing is to spend time using it. So as you read this book, if a question about the S language arises and you can't find the answer, try some of the examples on your own computer system; feel free to experiment! Many of the best techniques for using a language will be discovered on your own, just by trying out different ideas on different sets of data.

1.1 Invoking S

The command name for invoking S from the UNIX shell is generally the capital letter S. (Remember that UNIX is case-sensitive, so the lowercase letter s is treated differently from S). On some systems, a enhanced version of S, known as S-PLUS, will be available instead of, or in addition to, the usual version of S. This version can generally be accessed with the command Splus. (If you want to use command reediting, explained in Section 8.4, you should use the command Splus -e .) Under Windows, an "S-PLUS for Windows" icon can be selected to provide you with an interactive S-PLUS window. If none of these commands appears to be available, contact the person who installed S or S-PLUS on your computer for information on how to access the program.

1.2   Online Help

As you use S, you'll often need to get more information about some of the functions or data sets that are mentioned in this book. You can always find online help for all the built-in functions and data sets in S through the help function. Typing help(name) will display information on the named function or data set. You can also obtain help by preceding the name of a function or data set with a question mark (?). For example, typing

```
> ?help
```

would provide you with additional information on the help function. The help file entries for the functions discussed in this book will generally have information about options that are not mentioned here, so you should get in the habit of using the online help frequently. In addition, if new features or changes are added to the system, the online help will be updated, so it will always be the most up-to-date source of information.

A graphical interface to the help system is available when using S-PLUS under Windows or X Windows. Under Windows, the command help(), with no argument (but with the empty parentheses), will bring up this interface, which provides a scrolling list of topics to help you find the information you need. Once you've selected a topic, the names and brief descriptions of relevant functions are presented in a second scrolling list. When you choose a function from this list, a window displaying the help information you selected appears on the screen. A similar interface under X Windows (both Motif and OPEN LOOK) is available in S-PLUS through the command help.start. You must provide the name of the particular interface you are using through the gui argument of help.start; you can also set the gui option (Section 1.10) to the appropriate interface. For example, to view the selection of help topics under X Windows running OPEN LOOK, you can type

```
> help.start(gui="openlook")
```

or set the gui option and invoke help.start without arguments:

```
> options(gui="openlook")
> help.start()
```

The gui option can be automatically set each time you invoke S by including it in your own .First function; see Section 1.9 for details.

You can also get information about special symbols like <- or * from the online help system by enclosing them in quotes, or by selecting them from the list of topics in the graphical interfaces described above. For example, to get more information about the online help system accessed by typing the question mark, type

```
> help("?")
```

or

```
> ?"?"
```

If you're not sure which function will serve your purpose, you might start by looking at the topics presented through the graphical interface described above, or by looking at the section headings in Chapter 3, where many of the operators and functions in S have been organized according to functionality, each with a brief discussion of its use.

1.3   Data

Throughout this book, several example data sets will be used. These data sets should be available as part of the standard S distribution, so you can simply refer to them as they are used in the examples. These include the following:

1. **auto.stats** is a matrix of data concerning automobiles. The variables include fuel consumption (from the U.S. EPA) and repair, price, and dimensions (from *Consumer Reports* magazine). More details can be found by typing **help(auto)** from within S.

2. **saving.x** is a matrix of data containing the averages of five economic statistics for each of 50 different countries. Details can be found by typing **help(saving)** from within S.

3. **rain.nyc1** is a vector of New York City's rainfall in inches for the years from 1869 to 1957.

4. **state.x77** is a matrix containing various statistics about each state in the United States. Type **help(state)** for more information about these data.

In addition, you'll certainly want to use your own data to try out the ideas in this book. To create a vector containing a small number of values, you can use the **c** (for *combine*) function:

```
> mydata <- c(12,19,24,17,15)
```

For larger data sets, you'll most likely want to read your data from a file, with the S function **scan**. When you're using **scan** in its simplest form, each value in the file must be separated by one or more blanks or tabs. To read a vector, simply call **scan** with the filename in quotes as the argument, as in

```
> x <- scan("filename")
```

To read a matrix, insert the call to **scan** inside a call to the function **matrix**, along with the number of rows and/or columns in the matrix. Since S stores its matrices by columns, you must use the `byrow=T` argument if your data are stored by rows. For example, a 20 × 3 matrix of values could be read using

```
> mat <- matrix(scan("matrix.data"),ncol=3,byrow=T)
```

More information about reading your data into S can be found in Chapter 2.

1.4   The Basics of S

S is designed to be an interactive program, and you will probably use it that way most of the time. When you invoke S (usually with the command `S` or `Splus`), S will respond with a prompt, a greater-than sign (`>`), indicating that it is awaiting your input. Under Windows, S-PLUS provides a separate window that contains the prompt. Commands that you type in response to this prompt will be executed as soon as they are syntactically complete. This means that you don't need to restrict your commands to a single line. When S is waiting for you to complete a line—that is, to make it a syntactically acceptable S statement, it changes its prompt from the greater-than sign to a plus sign (`+`). For example, suppose you wish to calculate the mean of the New York City rainfall data set (`rain.nyc1`), and you forget to add the final parenthesis to the statement before you press Return. You can type the missing parenthesis on the next line after the `+` prompt:

```
> mean(rain.nyc1
+ )
[1] 42.31124
```

Once the statement is complete, S will execute it, and the result will appear on the screen. You can type multiple commands on a single line by separating them with a semicolon (`;`).

A wide variety of data and other information can be stored in S, and the various things that S creates and maintains are often referred to as objects. Objects in S can include numeric and character values as well as functions and collections of data of mixed types. Their names can be composed of letters, digits, and periods, but the first character of a name must be a letter. When deciding on names for the objects you use in an S session, you should also be aware that S is case-sensitive. Two variable names that differ only in case will refer to different objects.

One basic rule of S is that typing an object's name will display a representation of that object, unless an assignment is made to another object. For example, the S statement

```
> rain.nyc1
```

will display a printed representation of the object `rain.nyc1`, whereas the statement

```
> rain <- rain.nyc1
```

will set the value of the S object `rain` equal to that of `rain.nyc1` without displaying its value. The assignment operator `<-` sets the value of its left argument to that of its right argument. In words, this assignment operator is read as "gets"; for example, "rain gets rain.nyc1" would represent the statement shown above. (For convenience, the two symbols <- can be abbreviated as _; that is, the underscore symbol.)

If you issue a command to S, and forget to assign its returned value to another object, you'll simply see its printed representation and will not be able to refer to it or modify it. If you realize that you'd like to save the object right away, you can use the special system object called .Last.value to store the result, but be aware that the contents of the object .Last.value are volatile, and will change each time you enter a new expression. For example, suppose we wish to subtract the mean value of **rain.nyc1** from each element of **rain.nyc1**. If we type

```
> mean(rain.nyc1)
[1] 42.31124
```

the value of the mean will be displayed, but will not be stored in a variable for future use. We could save the value in the variable **mean.rain** by typing the command

```
> mean.rain <- .Last.value
```

We could now use this mean value in further computations; for example, to subtract the mean from each value, we could type

```
> new.rain <- rain.nyc1 - mean.rain
```

The function that informs S that you wish to end an S session is known as q, a mnemonic for *quit*. To quit an S session, type

```
> q()
```

The parentheses are required to inform S that you want to call the function, and not just to display a representation of that object. If you simply type q, you'll see a printed representation of the **quit** function; you can try it if you like!

1.5 A Sample S Session

In this section, an example of the use of S will be demonstrated, Many commands and concepts will be demonstrated without too much detail; the succeeding chapters of the book will help to fill in the details. If you wish to reproduce the graphs that are illustrated in the sessions, it will be necessary for you to choose an appropriate graphics device; see Section 4.1.

The data set to be studied is the **auto.stats** data set, which contains information about various automobiles, taken from *Consumer Reports* magazine and the U.S. EPA. We will look at the relationships among some of the variables measured for the automobiles. This data set is supplied with the standard S distribution, so typing its name will provide a display of the data set:

```
> auto.stats
                 Price Miles per gallon Repair (1978)
  Amc Concord    4099               22              3
   Amc Pacer     4749               17              3
  Amc Spirit     3799               22             NA
  Audi 5000      9690               17              5
   Audi Fox      6295               23              3
   BMW 320i      9735               25              4
Buick Century    4816               20              3
Buick Electra    7827               15              4
Buick Le Sabre   5788               18              3
  Buick Opel     4453               26             NA
  Buick Regal    5189               20              3
                        .  .  .
```

The actual display of the data set contains more variables; it has been abbreviated for convenience. Each observation is identified on the left-hand side by its name, and each column (variable) has a descriptive name. (The symbol NA in the display stands for "not available," and is used by S to represent missing values.) The labels used to display the data are known as dimnames; we can see their values by passing the auto.stats matrix to the dimnames function:

```
> dimnames(auto.stats)
[[1]]:
 [1] "Amc Concord"      "Amc Pacer"        "Amc Spirit"
 [4] "Audi 5000"        "Audi Fox"         "BMW 320i"
 [7] "Buick Century"    "Buick Electra"    "Buick Le Sabre"
[10] "Buick Opel"       "Buick Regal"      "Buick Riviera"
                        .  .  .
[[2]]:
 [1] "Price"            "Miles per gallon" "Repair (1978)"
 [4] "Repair (1977)"    "Headroom"         "Rear Seat"
 [7] "Trunk"            "Weight"           "Length"
[10] "Turning Circle"   "Displacement"     "Gear Ratio"
```

The second part of the dimnames display contains the variable names. We can use these names to access the variables we are interested in. For example, to learn more about the fuel economy of the cars, we could look at a histogram of the variable Miles per gallon. A histogram displays the range of a variable along the bottom of a graph, with bars of various heights indicating how many observations have values in each range. We can produce the histogram displayed in Figure 1.1 with the S command

```
> hist(auto.stats[,"Miles per gallon"])
```

The histogram shows that most of the cars have Miles.per.gallon values in the range of 15–20, and a few have values greater than 30. We could display the data for the cars with high mileage using the following statement:

Figure 1.1   Histogram of Miles per gallon

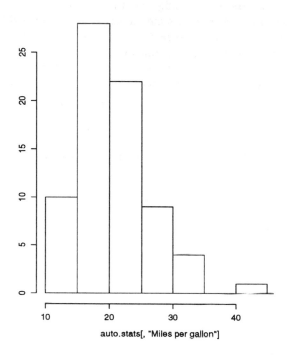

```
> auto.stats[auto.stats[,"Miles per gallon"] > 30,]
                Price Miles per gallon Repair (1978) Repair (1977)
   Datsun 210   4589               35             5             5
   Plym Champ   4425               34             5             4
       Subaru   3798               35             5             4
Toyota Corolla  3748               31             5             5
Volk Rabbit(d)  5397               41             5             4


                Headroom Rear Seat Trunk Weight Length Turning Circle
   Datsun 210      2.0      23.5      8   2020    165             32
   Plym Champ      2.5      23.0     11   1800    157             37
       Subaru      2.5      25.5     11   2050    164             36
Toyota Corolla     3.0      24.5      9   2200    165             35
Volk Rabbit(d)     3.0      25.5     15   2040    155             35
```

. . .

One way to display relationships between variables is with scatterplots—that is, plots containing one point for each observation (in this case, automobiles), with the x-axis representing one variable and the y-axis representing a second variable. For example, it seems natural that the weight of an automobile will affect its fuel economy, measured in miles per gallon. We can plot these two variables with a command such as this one:

```
> plot(auto.stats[,"Weight"],auto.stats[,"Miles per gallon"])
```

The plot is displayed in Figure 1.2. The relationship can be clearly seen: As the weight of the car increases, the economy, as measured in `Miles per gallon`, decreases.

Figure 1.2   Scatterplot of Automobile Weight and Fuel Economy

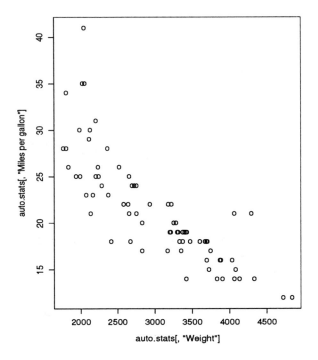

Transforming variables is S is easy: Simply assign the transformed value to a new variable name. S is smart about operating on all the values in a variable at once. For example, the previous scatterplot had some curvature in it, which might be removed by plotting the reciprocal of `Miles.per.gallon` against `Weight`. We can create a variable containing the reciprocal of `Miles.per.gallon`, say `rmpg`, with the statement

```
> rmpg <- 1 / auto.stats[,"Miles per gallon"]
```

We could now plot this variable against weight with a **plot** command similar to the previous one:

```
> plot(auto.stats[,"Weight"],rmpg)
```

The plot, shown in Figure 1.3, does seem to be straighter, except for the two outlying points. In Chapter 9, techniques will be shown for identifying points on a graph and for displaying informative labels.

Figure 1.3 Scatterplot of Automobile Weight and Reciprocal of Miles per Gallon

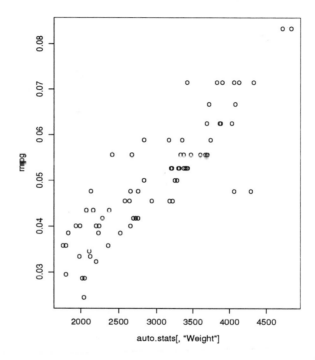

These simple examples should help to illustrate the kinds of things you can do using S, and to give you some simple ideas of ways in which you can manipulate and examine your data.

1.6 Using S with a Command File

Although most tasks in S are best performed interactively, there are times when it is useful to compose a file containing S commands and run them all at once, without any additional typing of commands. Such a file is variously known as a source file, a command file, or a batch file. If you have composed a file of S commands and wish to run them, you can use the **BATCH** option when invoking S or S-PLUS from the UNIX or MS-DOS command line, or under Windows using the File or Program Manager. When running the program in this way, you must provide two additional arguments: the names of the input and output files.

For example, if you had a file of S commands named **commands.s**, and you wished the S output from these commands to be placed in a file called **commands.out**, you would type one of the following commands at the UNIX or MS-DOS prompt:

```
S BATCH commands.s commands.out
```
or
```
Splus BATCH commands.s commands.out
```

When you are composing such a command file, it is useful to be able to annotate your program with comments—that is, lines of text that will remain in the file, but that S knows it should ignore. (Though S will accept comments in interactive mode, they are generally more useful when writing command files.) To have S ignore any text you choose to include in a command file, precede it with the pound sign (#). All text between the pound sign and the beginning of the next line will be ignored by S when it carries out your commands.

You can also execute S statements from a file while you are running S interactively, using the **source** function. From inside S, you could execute the commands in the file **s.cmds** by using the following call to **source**:

```
> source("s.cmds")
```

Keep in mind that no changes to any stored objects that might result from statements in a file of commands executed in this way will be carried out until after the successful execution of all the commands in the file. So if S encounters an error while running a set of commands through the **source** function, no changes to any S objects will take place. Also, the default behavior of automatic printing may be suppressed when you use the **source** function, so you may need to use the **print** function to make sure your output is displayed.

1.7 Interrupting S

If you type an S expression, and then realize that you don't really want it to execute, you can simply back up and type over the expression, provided you haven't yet pressed Return. If you have pressed Return, and the command has begun to execute, you can interrupt S's operation by pressing the Ctrl-C key combination under UNIX or the Ctrl-Break key combination under Windows. This action is known as an interrupt signal and is executed by holding down the Control key and the C or Break key at the same time. This will stop the currently executing command and return you to the S prompt.

1.8   Accessing Your Operating System from S

Sometimes it is necessary to directly execute a UNIX or MS-DOS command while you are in the middle of an S or S-PLUS session. For example, you may need to list the files in your current directory, print a file, or read your mail. When you are typing commands interactively (in response to the > prompt), you can precede any operating system command with an exclamation point (!), and S will immediately pass the command to the operating system. For example, on a UNIX workstation, to read your mail from inside of S, type

```
> !mail
```

When the `mail` command is completed, control is returned back to S.

To achieve the same effect when running commands in a source file, you can enclose the operating system command in quotes and pass it to the **unix** or **dos** function, with the optional **output=F** argument. For example, under MS-DOS or Windows, to print the date on which a set of commands is executed from a source file, you can include a line like this at the beginning of the file:

```
dos("date")
```

The output from the MS-DOS command **date** will be included as part of the output from your S session, either displayed on the terminal or redirected to a file.

1.9   Initialization and Wrapup

As you become more familiar with S, you may find that there are certain things you like to do each time you start or end an S session. For example, using S-PLUS, you may want to open the interactive help window each time you start up a session. To allow you to do this, there are two special S objects, `.First` and `.Last`, which are called each time S starts or finishes, respectively. These objects should be functions that accept no arguments. (A complete discussion of functions can be found in Chapter 6; for now, the simple examples below will suffice.) For example, you could define `.First` as follows:

```
> .First <- function()objects()
```

to list the names of all the objects in your S data directory each time you start an S session. (See Section 2.4 for more information on the **objects** function). Similarly, you could print a farewell message at the end of each S session by defining `.Last` as

```
> .Last <- function()cat("Goodbye for now!\n")
```

The function `cat` displays its arguments on the computer screen. Examples of more complicated functions will appear in later chapters.

1.10 Options

The behavior of S can be modified in a number of ways by changing the values of a variety of options. You can view the current settings of all the options of the system by typing `options()`; to access just a single value, use `options("optname")`, where `optname` is the name of the option in question. Table 1.10 lists the types and functions of a number of the more common options for the S system; for a complete description of the options available on your system, consult the online help (`help(options)`).

Table 1.1 Common Options for the S System

Printing

Option	Type	Effect	Default
echo	logical	expressions are displayed before they are evaluated	FALSE
width	integer	number of columns for printing	80
length	integer	number of lines for printing	48
digits	integer	number of significant digits to be displayed	7

Limits

Option	Type	Effect	Default
check	logical	perform internal memory checks	FALSE
memory	integer	number of bytes of internal memory	50000000
object.size	integer	maximum number of bytes for generated objects	5000000
audit.size	integer	maximum allowable audit file size in bytes	500000

Display

Option	Type	Effect	Default
prompt	character	string to be used as primary prompt	>
continue	character	string to be used as secondary prompt	+
error	function	function to be called when an error occurs	dump.calls
editor	character	default text editor	ed
warn	integer	level of warnings displayed	0
gui	string	graphical user interface (S-PLUS only)	none

Options that are logical values can be either true (`TRUE` or `T`) or false (`FALSE` or `F`).

To change the value of an option, use a statement of the form `options(name = value, ...)`. For example, to change the primary prompt to a question mark (`?`) and the printing width to 72 characters, you could type

```
> options(prompt="?",width=72)
```

Keep in mind that changes to options are in effect during the current S session only, and will need to be reset the next time you start an S session. The `.First` function, mentioned in the preceding section, is a good place to install any changes to options that you would like to use every time you invoke S.

Exercises_____

1. One of the most important tasks in using S is reading data from external sources into S objects, so that you can perform calculations and produce graphics. If you have a data set that you're interested in, try reading it into S using the **scan** and **matrix** commands as briefly described in Section 1.3.

   If you don't have a file of data to read, you can use the **write** command to produce one. For example, to write out the matrix **state.x77** by rows, you can use the expression

   ```
   > write(t(state.x77),file="state.out",n=8)
   ```

   Use the above statement to create a file with the state data in it, then use the **scan** and **matrix** commands to read it back into S. Why was the **t** (transpose) command used in the write statement? What happens if you don't transpose the matrix before writing?

2. Continue the sample session using the **auto.stats** data set by experimenting with some of the graphical techniques described in Chapter 4. For example, try:

   a. using the **lm** and **abline** functions to add a regression line to the scatterplot.

   b. using the **identify** function to identify the outliers on the scatterplot.

   c. using the **split** and **boxplot** functions to make boxplots of gas consumption for cars of different weights.

2 Data in S

2.1 Information About Objects

Anything that exists inside of S is referred to as an object. As mentioned previously, typing the name of an object from inside S will display a printed representation of that object. Internal to S, however, additional information is usually stored along with the object. For example, every object in S has a length and a mode. The modes of data in S include numeric, character, logical, and complex, and many S functions are clever about the way they deal with objects of different modes. Often an error message will refer to the mode of an object, because some operations are inappropriate for certain modes of data. For example, mathematical operations are inappropriate for character data. You can use the functions **length** and **mode** to display this information for the objects you are working with.

In addition to this basic information, additional information can be stored as attributes, which are discussed in Section 2.9.1.

2.2 Reading Data into S

To create an object within S, you generally assign the output of a function to the name of the object in question. For example, the **c** (mnemonic for *combine*) function of S combines similar objects into a vector. So to create a vector called **y** containing the numbers 4, 7, and 12, we could use the S statements

```
> y <- c(4,7,12)
> y
[1]  4  7 12
```

The same function will work with character vectors.

```
> labs <- c("Fred","Joe","Tim","Earl")
> labs
[1] "Fred" "Joe"  "Tim"  "Earl"
```

The c function is fine if you have only a few values to enter, but a more useful function for getting data into S is scan. This function allows you to read either data from a file or data entered from the keyboard; the data values you read are separated by one or more blanks, tabs, or newlines. In computer jargon, the data values must be separated by white space. You can also specify a separator other than white space by using the sep= argument described below. When scan is called with no argument, it prompts for data at the keyboard. For example, the statements given below will create a vector called z containing the listed numbers. Input is completed when a blank line is typed.

```
> z <- scan()
1: 3.1 5.0 9.32 12
5: 19 22.453
7: 14
8:
> z
[1]   3.100   5.000   9.320 12.000 19.000 22.453 14.000
```

The numbers on the left, before each colon, show which element of the vector should be entered next, and can serve as a guide indicating how many elements have already been entered. The final line resulted from pressing Return with no other input after the last number was read in. By default, scan expects numeric values. If you wish to read character values, you can inform scan through the use of the optional what= argument, by providing a pattern of the type of data you will be passing. For character data, this can simply be an empty quoted string ("") If there are embedded blanks in a character value, the value should be enclosed in double quotes. You can also specify a separator other than the default of white space by using the sep= argument. So to read in a series of automobile names using scan, you might use the following:

```
> cars <- scan(what="",sep=":")
1: Honda Civic:Ford Escort:Toyota Celica:
4: Chevrolet Chevette:Pontiac Firebird
6:
> cars
[1] "Honda Civic"        "Ford Escort"        "Toyota Celica"
[4] "Chevrolet Chevette" "Pontiac Firebird"
```

Without the sep=":" argument, character strings with embedded blanks would have to be surrounded by quotes in order to be read properly:

```
> cars <- scan(what="")
1: "Honda Civic" "Ford Escort" "Toyota Celica"
4: "Chevrolet Chevette" "Pontiac Firebird"
6:
> cars
[1] "Honda Civic"        "Ford Escort"        "Toyota Celica"
[4] "Chevrolet Chevette" "Pontiac Firebird"
```

The quotes are not stored as part of the actual character variable's value. You can display character values without quotes by passing the argument **quote=FALSE** to the **print** function:

```
> print(cars,quote=FALSE)
[1] Honda Civic        Ford Escort        Toyota Celica
[4] Chevrolet Chevette Pontiac Firebird
```

Sometimes it's helpful to have a descriptive character string associated with each of the elements of a vector. For example, we might have a vector of length 3, containing a person's age, height, and weight. If we were to print the vector, it would be desirable to label each element accordingly. Such labels can be stored through assignment to the **names** function. Continuing the example:

```
> info <- c(25,69,162)
> info
[1]  25  69 162
> names(info) <- c("Age","Height","Weight")
> info
 Age Height Weight
  25     69    162
```

After assigning the values through the **names** function, the labels will be used to identify the elements whenever the vector is printed. In Section 5.1.1, it will be seen that these names can also be used to access elements of the vector directly.

If you pass a filename (in quotes) as the first argument to **scan**, it will take its input from the named file instead of from the keyboard. The same rules about separators apply when you are reading input from a file. If your data are not in this format—for example, if your input contains data in fixed column fields— the function **make.fields** may be useful in rewriting your data to make it acceptable for **scan**. Some additional features of **scan** are documented in Section 2.9.4.

Because character strings in S-PLUS use backslashes (\) to signal special characters such as newline (\n) or tab (\t), a single backslash in a filename will not be properly recognized by S-PLUS running under Windows or MS-DOS. You can either include two backslashes or replace the backslash with a forward slash (/); in either case, S-PLUS will interpret it as a backslash and pass it to the operating system. For example, suppose you wish to read a vector of data from an MS-DOS file called `a:\data\rocket.dat`. You could use either of the following statements to access the file for use with the `scan` command:

```
> rocket <- scan("a:\\data\\rocket.dat")
```
or
```
> rocket <- scan("a:/data/rocket.dat")
```

2.3 Generating Data in S

Though some operations in S will require you to read data from an outside source, as discussed in the previous section, there are times when you need to create data within S without relying on any outside source. For example, you may wish to produce a plot in which one of the variables consists of the integers from 1 to 100. Analysis of designed experiments often requires that treatments be specified in a repeatable pattern. When you are testing out a new statistical technique, it may be desirable to use a set of numbers that arises from a specified distribution, without having to generate those numbers from an outside source. Tasks such as these are easily handled in S using a variety of built-in functions and operators.

The simplest operator for generating data in S is the sequence operator, represented in S as a colon (`:`). The sequence operator generates a vector of integer values within a given range in increments of either 1 or -1. For example:

```
> nums <- 1:10
> nums
 [1]  1  2  3  4  5  6  7  8  9 10
> nums1 <- 5:-4
> nums1
 [1]  5  4  3  2  1  0 -1 -2 -3 -4
```

Since the sequence operator operates only in steps of 1 or -1, you may need to modify its output if you wish to generate a sequence from, say, 0 to 1 in steps of 0.1. One way to do this is simply to divide the sequence from 0 to 10 by 10:

```
> 0:10/10
 [1] 0.0 0.1 0.2 0.3 0.4 0.5 0.6 0.7 0.8 0.9 1.0
```

Another alternative is to use the `seq` function, which lets you specify arguments for `to=`, `from=`, and `by=`. The sequence described above could be generated using the following statement:

```
> seq(from=0,to=1,by=.1)
 [1] 0.0 0.1 0.2 0.3 0.4 0.5 0.6 0.7 0.8 0.9 1.0
```

As an alternative to the `by=` argument, you can specify the length of the desired sequence with the `length=` argument; only one of these two arguments should be used. The results in the above example could be obtained using the following statement:

```
> seq(from=0,to=1,len=11)
 [1] 0.0 0.1 0.2 0.3 0.4 0.5 0.6 0.7 0.8 0.9 1.0
```
Note that the length argument's name was shortened to len. Except in special circumstances (see Section 6.11.1), you can abbreviate an argument name, provided that there is no other argument with which it could be confused.

Another useful argument to the seq function is along=, which will generate a sequence with as many elements as the object of the along= argument. This is especially valuable when you have an object of length zero, as illustrated by the following statements:

```
> tst <- NULL
> length(tst)
 [1] 0
> 1:length(tst)
 [1] 1 0
> seq(along=tst)
numeric(0)
```
The built-in value NULL represents an empty object—that is, an object with length of zero—so the : operator mistakenly generates a sequence from 1 to 0. The seq function is smarter, and correctly returns a numeric object of length zero.

To generate repeated values, you can use the rep function. For example, in designed experiments it is often necessary to generate repeated occurrences of a sequence of integers. To create a vector containing, say, the integers from 1 to 5 repeated three times, we could use the following statement:

```
> rep(1:5,3)
 [1] 1 2 3 4 5 1 2 3 4 5 1 2 3 4 5
```
As with most data manipulation functions in S, the first argument to rep is not limited to a sequence of integers, or even to numeric mode. If the levels of a designed experiment were represented by the letters A, B, and C, a vector of repeated values could be generated as follows:

```
> rep(c("A","B","C"),4)
 [1] "A" "B" "C" "A" "B" "C" "A" "B" "C" "A" "B" "C"
```
You can also provide a vector of values representing the number of times the arguments to rep should be repeated. For example, to produce a vector with five 1s, four 2s, three 3s, two 4s, and one 5, we could use the expression

```
> rep(1:5,5:1)
 [1] 1 1 1 1 1 2 2 2 2 3 3 3 4 4 5
```
If you use a vector as the second argument to rep, it must be of the same length as the first argument.

Another simple and useful function is rev, which takes a vector as its argument and returns a vector of the same length with its elements in reverse order.

A wide variety of random number generators are available in S. Though a complete discussion of these capabilities can be found in Section 3.10.4, two of these functions deserve mention now: **rnorm**, which generates normally distributed numbers, and **runif**, which generates uniformly distributed numbers. The important thing to note about these functions is that they will generate data without the regularity of data produced by the **seq** and **rep** functions, so they are useful for providing a wider range of test data than those functions can provide.

When given a single number as an argument, both **rnorm** and **runif** will produce a vector whose length is equal to the argument given, with the parameters of the distribution set to default values. Values generated from random number generators are often more useful for testing new programming techniques or functions than data with regular structure, such as vectors produced by **rep** or **seq**.

For example[1], we could generate a vector of 100 normally distributed random numbers with the function **rnorm**, and then display a stem and leaf diagram of the data with the statements below. (A stem and leaf diagram is like a histogram turned sideways.)

```
> x.norm <- rnorm(100)
> stem(x.norm)

N = 100    Median = 0.2505125
Quartiles = -0.577683, 0.989959

Decimal point is at the colon

  -2 : 321
  -1 : 877655
  -1 : 4322100
  -0 : 99887777655
  -0 : 4333332221000
   0 : 00111112223333444444
   0 : 56666777778
   1 : 00000001112223334
   1 : 555669
   2 : 00024
   2 : 6
```

Similarly, we could generate 100 uniform random numbers using **runif**. Notice the difference in the shapes of the two distributions:

---

[1] If you try this example on your computer, you'll get similar but not identical results. This is because the sequence of random numbers changes each time a function such as rnorm is called. See Section 3.10.4 for more details.

```
> x.unif <- runif(100)
> stem(x.unif)
```

```
N = 100    Median = 0.4698843
Quartiles = 0.2788418, 0.737506
```

```
Decimal point is 1 place to the left of the colon
```

```
0 : 13456789
1 : 1234445689
2 : 002247789
3 : 233346677
4 : 000011234556777
5 : 11134456667888
6 : 123448
7 : 0123567778899
8 : 0345559
9 : 001236689
```

2.4 How S Finds Data

If you create an object called x in an S session, it is not surprising that when you refer to that object later in the session, it has the value you expected. But what happens if you exit S and reinvoke it at some later time? As it turns out, S will "remember" the values of all the objects you create from one session to the next, by storing them in a special directory. S will first look in the current directory for a subdirectory named .Data; if such a directory is not found, it will look for a subdirectory called .Data in your home directory. (In S-PLUS running under MS-DOS, the directories are named _Data instead of .Data.) If neither of these directories exists or if you do not have write access to these directories, S will inform you that no appropriate data directories exist and will prompt you for the name of the directory to be created with a message such as the following (the % symbol stands for the UNIX prompt from your computer):

```
% S
None of the following directories exist with
read/write/search permission:
.Data
/usr/home/.Data
```

```
To create one or fix its permissions, type its name;
Otherwise, type 'q' or a blank line to exit
```

If you have called S from your home directory, the correct response will usually be .Data. If you type .Data after seeing this message, the appropriate directory will be created. Under S-PLUS, the .Data or _Data directory is automatically created and you won't see any warning message.

Once the directory named .Data exists in your home directory, you will not see the message again. Alternatively, you can use the UNIX or MS-DOS command mkdir to create a .Data (under UNIX) or _Data (under MS-DOS) subdirectory in any directory of your choosing. Since S will first look for such a directory in the current working directory before looking in your home directory, this provides a mechanism for organizing the objects you create by project, namely creating separate data directories in a different directory for each project. Then, if you set your working directory to one of the directories containing a .Data subdirectory before invoking S, it will use that .Data subdirectory to store the objects you create during that session. The UNIX or MS-DOS command cd can be used to change the value of your working directory, or you can choose the Properties selection from the Program Manager under Windows.

Your own local data directory is not the only directory that S searches when looking for an object, although it is the only directory that S will use to store objects that you create. You can see the directories that S is searching by using the S function search. The results of a call to search might yield output such as the following:

```
> search()
[1] ".Data"
[2] "/home/s/s/.Functions"
[3] "/home/s/s/.Datasets"
```

The first directory, known as the working database, is your local data directory described above. New objects are stored and modifications to existing objects take place in the working database. The next two directories are the S system function and data set directories. In addition, you may see other directories from libraries, which are special directories containing related S objects. Libraries will be discussed in more detail in Section 2.8. You can add directories to the search list with the function attach, which is discussed in more detail in Section 2.9.2. You can see the names of all the objects in any of the directories on your search list by using the function objects. With no arguments, objects returns the names of all the S objects in the working directory, which is usually your .Data directory. For example, if you have been testing the examples in this chapter, a call to objects might result in the following:

```
> objects()
[1] ".Last.value"  ".Random.seed" "cars"     "info"
[5] "labs"         "nums"         "nums1"    "x.norm"
[9] "x.unif"       "y"            "z"
```

To list the objects in another directory in your search path, you can pass an argument of the form where=n, where n is the location of the directory of interest in your search list. You can use the search function to find the location of the directory of interest. For example, to display a list of the objects found in the system function directory, we could first find its location using `search`, then pass this information to the `objects` function:

```
> objects(where=2)
```

because the system function directory is the second directory on the search list, as indicated by the call to `search` that was displayed above.

The `objects` function can be instructed to restrict its search for objects to those whose names contain a particular pattern of characters, by passing the pattern to `objects` with the `pattern` argument. For example, you could find all objects in the system data sets directory whose names contain the pattern *state* with the command

```
> objects(pattern="state",where=3)
[1] "city.state"     "gr.state"        "state.abb"
[4] "state.center"   "state.division" "state.name"
[7] "state.region"   "state.x77"
```

In S-PLUS, the function `objects.summary` can be used to produce a table that lists, along with the name of each selected object, its mode, class, size, and date of last modification. The `pattern` and `where` arguments described above are also valid for `objects.summary`, along with a variety of other optional arguments to control the information produced.

When S tries to find an object, it looks in the directories in your search path, in order, seeking an object with the name of the one it is looking for, and stops searching after it finds one.[2] One consequence of this is that if you create an object in your data directory with the same name as one in a system directory, S may get confused when it tries to find the object in question. For example, the S function `c` is used by many other functions in S. What happens if you create a numeric object called `c` in your local data directory? When S looks for the object called `c`, which it expects to be a function, it will encounter the numeric object in your data directory first. Since this object is not a function, S continues to look for an appropriate function in the other directories in your search path; it will eventually find one in the system `.Functions` directory. Because it has found another object with the same name, however, it will issue a message such as the following:

```
Warning messages:
  Looking for object "c" of mode "function", ignored one of
  mode "numeric"
```

This warning simply means that your use of a numeric object with the same name as a system function is confusing the S interpreter. Usually the best solution is to assign the object to some other name, then remove the offending object with the S function `rm`. For example, suppose you saw the above message regarding the object called `c`, and realized that you had stored a set of numbers in the

---

[2]Inside functions, objects are created in a local evaluation frame, so that names of objects inside functions do not interfere with names of objects stored in directories. The mechanism for finding objects inside functions is discussed in more detail in Section 6.9.

object. You could save the object under the name of `cc` (which will not interfere with any system objects), and remove the object named `c` with the following statements:

```
> cc <- c
> rm(c)
```

See Section 6.6 for more information about object name conflicts.

2.5 Organizing Data

In earlier sections of this chapter, it was seen that when several data values are combined, the result is a vector of values. For example, using the function `c` or `scan` with more than one value creates a vector of those values. Although vectors are a logical extension of single values, there are many cases in which thinking of your data as having additional structure can be very useful. Many mathematical operations are simplified by viewing data as a matrix—that is, a two-dimensional array—and viewing a collection of data as a set of variables measured on a number of observations can expedite most statistical analyses. The following subsections present the ways in which data can be organized and stored within S.

2.5.1 Organizing Data: Lists

In order to manage large amounts of information in an efficient way, it is often necessary to store different types of data within the same S object. The results of a statistical analysis, for example, might contain a vector of parameter estimates, a vector of residuals, labels for variables and/or observations, or even a function to plot a graph or calculate a quantity of interest. An object of this type, composed of more than one S object, is known as nonatomic or recursive in S, and is characterized as a list. A list is the basic means of storing a collection of data in S when the modes and/or the lengths of the objects are not the same. There are few restrictions on the types of objects that can be stored in a list. One very important use of lists is in returning information from functions. For example, the statistical modeling functions, such as `lm` and `aov`, return information through lists, and many of the other S functions (such as `print` and `summary`) have special methods for handling such lists. (Such techniques are known as object-oriented programming, and are discussed in Section 7.6.3.)

You can create a list in S with the function called `list` by supplying as arguments each of the objects you want stored in the list. To make access to these objects easier, it's usually a good idea to name the elements of the list. For example, if we wished to store some of the objects created by the examples in the previous sections in a list called `examples`, we could use the following statement:

```
> examples <- list(labels=labs,cars=cars,info=info)
```
Within the list, these objects would be known as **labels**, **cars**, and **info**, respectively. A list's length is defined as the number of objects in the list, so the **examples** list defined above would have a length of 3. To refer to the elements of a list, use the dollar-sign operator (**$**), as in **examples$cars**.

2.5.2   Organizing Data: Matrices

When all of your data are of the same mode, an array is often the most effective way to store them. An array is a collection of data, all of the same mode, which is indexed by one or more subscripts. Subscripts in S follow the name of the object to be subscripted, and are enclosed in square brackets (**[ ]**). The vectors discussed previously can be thought of as one-dimensional arrays, because each element of an array can be indexed by a single number. Given a vector **x**, for example, the first element would be referred to as **x[1]**, the second as **x[2]**, and so on. Matrices, or two-dimensional arrays, are the most commonly used, but higher-dimensioned arrays are also possible. Since matrices are so commonly used, we'll focus on them.

Not surprisingly, the basic function for creating a matrix is called **matrix**. After reading a vector of data into S, you can convert it to a matrix by calling the **matrix** function with additional arguments specifying the number of rows (first dimension) and/or columns (second dimension) of the matrix you wish to form. If you leave out one or the other, the **matrix** function will determine the missing argument by dividing the length of the data being converted to a matrix by the dimension that was supplied in the argument list. The matrix will then be formed, possibly repeating values from the data to fill out the matrix to the specified dimensions. For example, if we pass a vector of length 5 to the matrix function, and ask for a matrix with five rows and three columns, it will repeat the five values in each row to comply with the request:

```
> rowdata <- c(3,7,9,12,8)
> my.matrix <- matrix(rowdata,nrow=5,ncol=3)
> my.matrix
     [,1] [,2] [,3]
[1,]    3    3    3
[2,]    7    7    7
[3,]    9    9    9
[4,]   12   12   12
[5,]    8    8    8
```
This feature is especially useful when a scalar value is passed to the matrix function; in that case, **matrix** will return a matrix of the specified dimensions containing the scalar in every position. To create a matrix with six rows and four columns, containing a zero in every position, we could use the following statements:

```
> zeroes <- matrix(0,nrow=6,ncol=4)
> zeroes
      [,1] [,2] [,3] [,4]
[1,]    0    0    0    0
[2,]    0    0    0    0
[3,]    0    0    0    0
[4,]    0    0    0    0
[5,]    0    0    0    0
[6,]    0    0    0    0
```

You can determine the size of an existing matrix (or other array) by using the function dim. dim accepts a single argument, the object in question, and returns either NULL, if the object is not an array, or a vector whose length is equal to the number of dimensions of the array and whose elements contain the extent of each dimension. For example, when dim is called with a matrix, it returns a vector whose first element is the number of rows in the matrix and whose second element is the number of columns. For convenience, two S functions also exist that will return the number of rows or columns in a matrix or other array, namely the functions nrow and ncol. Like dim, they return NULL if the object in question is not an array. You can see the behavior of these functions when applied to vectors and matrices in the following examples, using objects created earlier in this section:

```
> dim(rowdata)
NULL
> nrow(rowdata)
NULL
> dim(my.matrix)
[1] 5 3
> nrow(my.matrix)
[1] 5
> ncol(my.matrix)
[1] 3
```

Though the inner workings of S are generally of no interest to most users, one important fact about the way matrices are stored in S must be kept in mind when using the matrix function: Matrices are stored within S by columns. This means that if you supply a vector of values to the matrix function, it will assume that the vector contains the first column of the matrix immediately followed by the next column, and so on. For example, if we form a matrix of two columns from the numbers 1 through 10, we obtain the following:

```
> matrix(1:10,ncol=2)
      [,1] [,2]
 [1,]    1    6
 [2,]    2    7
 [3,]    3    8
 [4,]    4    9
 [5,]    5   10
```

This default behavior can be overridden using the **byrow=** argument to the **matrix** function. When **byrow** is set to **TRUE**, S will assume that the matrix is stored by rows instead of by columns, as in the following example:

```
> matrix(1:10,ncol=2,byrow=T)
      [,1] [,2]
 [1,]    1    2
 [2,]    3    4
 [3,]    5    6
 [4,]    7    8
 [5,]    9   10
```

The **byrow=T** argument is especially useful when you are reading a matrix from a file, because data in a file are generally stored in a row-by-row (observation-by-observation) fashion. Because the output of an S function can be used as input to another S function, a matrix can be read from a file by a statement similar to the following, assuming there are three columns in the matrix:

```
> mymat <- matrix(scan("datafile"),ncol=3,byrow=T)
```

Because the rows and columns of a matrix often represent observations and variables, respectively, which may have meaningful names of their own, S provides a mechanism for storing these names transparently along with the data values in a matrix. These names are known as dimnames; each consists of a list of length 2. The first element of the dimnames list is a character vector whose length is equal to the number of rows in the matrix; the second element is a character vector whose length is equal to the number of columns in the matrix. You can create dimnames either by passing the list to the **matrix** function with the optional **dimnames** argument or by using the **dimnames** function on the left-hand side of an assignment statement. (Also see the next subsection, on data frames, if you have a data set that fits this description.) Suppose we create a data matrix called **kids**, whose first column contains heights of children and whose second column contains the corresponding weights of the children. By including a **dimnames** argument to the matrix function when the matrix is created, we can provide labels for the rows and columns of the matrix as follows:

```
> kids <- matrix(scan(),ncol=2,byrow=T,dimnames=
+ list(c("Tom","Fred","Sue","Donna","Lee","Jack"),
+ c("Height","Weight")))
1: 42 66 41 62 40 62 43 71 43 79 42 67
13:
```

```
> kids
        Height  Weight
    Tom     42      66
   Fred     41      62
    Sue     40      62
  Donna     43      71
    Lee     43      79
   Jack     42      67
```

Alternatively, the matrix could be created in the usual fashion, and the dimnames assigned after it is created, using the `dimnames` function on the left-hand side of the assignment statement:

```
> kids <- matrix(scan(),ncol=2,byrow=T)
1: 42 66 41 62 40 62 43 71 43 79 42 67
13:
> dimnames(kids)<-list(c("Tom","Fred","Sue","Donna","Lee","Jack"),
+                       c("Height","Weight"))
```

If you have only row or column labels, but not both, you can replace one of the two vectors in the dimnames list with `NULL`. In addition to displaying the dimnames of a matrix when it is printed, many S functions will use the dimnames when producing printed output or graphs. The dimnames can also be used to access data directly, as explained in Section 5.1.1.

2.5.3   Organizing Data: Data Frames

A data frame is a cross between a list and a matrix, combining the best properties of both, including allowing easier access to variables (columns) within a data set by reference to their names. Each component of a data frame must be of the same length (like the rows or columns of a matrix), but the modes of the different components need not be the same (as in a list). Thus, numeric and character data may be stored in the same object by using a data frame, but functions such as `dim`, `dimnames`, `nrow`, and `ncol` all work on data frames as if they were matrices. It is helpful to think of a data frame as a rectangular array in which the rows represent observations and the columns represent variables; that is, each component of the data frame represents the values of a particular variable in the data set for each of the observations. Therefore, if you view your data as a set of observations on a variety of variables, a data frame will be the best way to store them. This is especially true if you will be performing statistical modeling on your data, because the statistical modeling functions have special provisions for working with data frames.

To allow easy access to data in data frames through variable names, the variable names of a data frame cannot contain blanks, parentheses ((or)), or square brackets ([or]). (There are no such restriction on the row (observation) labels, however.)

   Because of their close relationships to matrices and lists, data frames can
simply be converted from either one through the use of the `data.frame` function.
For example, the matrix `state.x77`, included in the S distribution, is a matrix
containing information about the 50 U.S. states, with the states' names serving
as the first component of the dimnames— that is, the row labels of the matrix.
(For more information about this data set, use the S command `help(state)`.)
To convert this matrix into a data frame called `state`, we could use the following
command:

```
> state <- data.frame(state.x77)
```

Each column of `state.x77` would become a component of the data frame `state`.
The dimnames values associated with `state.x77` will be used to provide row
and/or column names. Because of the restrictions on variable names described
above, embedded blanks and special characters in the column dimnames of the
`state.x77` matrix are replaced with periods (`.`). Thus, instead of referring to
`state.x77[,"Life Exp"]`, we can use the expression `state$Life.Exp`.

   One of the major advantages of using data frames is that the statistical
modeling functions described in Chapter 10 will all accept an optional argument
called `data`, which allows you to supply a data frame to the functions and then
refer to the variables in the data frame by name. In addition, a data frame may
be attached to the search path described in Section 2.4, allowing the names given
to the variables in the data frame to be used without the name of the data frame
itself. If we attached the `state` data frame to the search list, we could refer to
the variables within as `Life.Exp`, `Income`, and so on, without having to specify
that these variables come from the `state` data frame:

```
> attach(state,1)
> mlife <- mean(Life.Exp)
[1] 70.8786
> maxinc <- max(Income)
> maxinc
[1] 6315
```

The optional second argument to `attach` ensures that the `state` data frame will
be attached in the first position on the list, so that the names of the variables will
be properly resolved even if there are objects of the same name in your `.Data`
directory. Once the data frame is attached, the variables in the data frame can
be referred to directly in any S expression. This scheme will cause no problems
if the variables in the data frame are not modified. (A discussion of attaching
and detaching data frames when values are to be modified will be presented in
Section 2.9.2.)

   Data frames can be formed directly by combining components in the
`data.frame` function. For example, we could create a data frame containing
information about the numbers of games won and lost for some football teams
with the following expressions:

```
> team <- c("Falcons","Bears","Eagles","Dolphins")
> wins <- c(5,11,10,12)
```

```
> losses <- c(11,5,6,4)
> football <- data.frame(team,wins,losses)
> football
        team wins losses
1  Falcons    5    11
2    Bears   11     5
3   Eagles   10     6
4 Dolphins   12     4
```

The names for the variables in the data frame come from the names of the objects from which the data frame was formed. You can specify variable names for the data frame when it is created, by naming the arguments to `data.frame`. Row labels can be provided through the `row.names` argument to `data.frame`:

```
> fball <- data.frame(Team=team,Wins=wins,Losses=losses,
+        row.names=c("Atlanta","Chicago","Philadelphia","Miami"))
> fball
                Team Wins Losses
     Atlanta  Falcons    5    11
     Chicago    Bears   11     5
Philadelphia   Eagles   10     6
       Miami Dolphins   12     4
```

We can now refer to variables as `fball$Wins` and `fball$Losses` in any S expression, or, alternatively, attach the `fball` data frame and refer simply to `Wins` and `Losses`.

Finally, the function **read.table** will read the contents of an external file into a data frame. This function is similar to `scan`, but has several extensions to provide extra flexibility. As with `scan`, you can specify a separator other than the default of whitespace by providing a character to the `sep=` argument of `read.table`, but you can alternatively provide a vector of column numbers from which `read.table` should start reading the variables for each observation. This allows `read.table` to read data stored in fixed fields directly, without preprocessing by operating system commands. One limitation of `read.table` is that all of the data from a single observation must appear on the same line in the file. However, unlike `scan`, `read.table` does not require you to inform S of the nature of your variables, because it will determine whether variables are numeric or character as it reads the data. For added convenience, when **header = TRUE** is specified, `read.table` interprets the entries in the first line of your file as variable names to be used when the data frame is formed. These variable names must be separated from each other in the same fashion as the data itself. Alternatively, the two optional arguments **row.names** and **col.names** can be used to provide observation and variable names, respectively. The **row.names** argument can be either a vector of row names or a single number or character string specifying which variable in the data frame should be used to construct row names. If no row names are given, `read.table` uses the first nonnumeric variable it encounters in the data; if no such variable is present, the row names default to character representations of the observation numbers. The `col.names` argument is a vector of character values to be used as column names if no headers are provided. By default, `read.table` will construct column names by appending a number to the letter `V`.

To illustrate the use of `read.table` with fixed format data, consider a data file called `fixed.data`, containing the following lines:

```
12First 19 2134
15Second29 4213
18Third 31 2619
```

Assuming that columns 12 through 15 represent two data values, each occupying two columns, the following statement will read the data and place them in a data frame:

```
> read.table("fixed.data",sep=c(1,3,9,12,14))
        V1 V3 V4 V5
  First 12 19 21 34
 Second 15 29 42 13
  Third 18 31 26 19
```

The use of `read.table` with free-format data can be illustrated by considering an example similar to the preceding one. Suppose a file called `f.data` consists of the following:

```
Names  X Y labels
Fred 15 2 First
Tim 24 4 Second
Sue 19 8 Third
John 7 3 Fourth
```

We could read these values directly into a data frame using the statement

```
> read.table("f.data",header=T,row.names="labels")
          Names  X Y
   First  Fred 15 2
  Second   Tim 24 4
   Third   Sue 19 8
  Fourth  John  7 3
```

If you have problems reading data using `read.table`, the function `count.fields` may be useful. It reports the number of fields on each line of a file whose name is passed to `count.fields` as the first argument. An optional `sep` argument can also be provided if a field separator other than blanks is used.

2.6 Time Series

Many collections of data are actually measurements of the same thing taken repeatedly over a period of time. For example, a patient's pulse rate might be measured every hour, or the ozone concentration in a city's air might be measured every day. A collection of data such as this is often called a time series. S has special facilities for storing time series, as well as functions that take into account the time-series nature of data for calculations and displays.

To create a time series in S from a vector of data, use the function `ts`. The first argument to `ts` is the data that are to be made into a time series. The next argument, named `start`, gives the time of the first observation in the data set. If you were using annual data (one observation per year), you would use the first year's value for `start`; for monthly data, you could express the starting date either as a year followed by + and a fraction, or as a vector of length 2 consisting of the year followed by the number of the month. For example, November 1990 could be expressed either as `1990 + 11/12` = `1990.917` or as `c(1990,11)`. The next argument to `ts` is frequency, which is the number of observations per year; for example, `frequency = 1` for annual data, or `12` for monthly data. As an alternative to the `frequency` argument, you can use the `end=` argument to `ts`, which consists of the final date of the time series in the same format as the `start` argument. You can use a time scale other than years to describe a time series. For example, hourly data could be presented starting at 1 and with a frequency of 24. Operators such as addition and subtraction will act appropriately on time series; that is, they will carry out the specified operation only on observations from the same time period. For example, the data set `corn.rain` represents inches of rainfall in six states for the years 1890 through 1927, whereas the data set `rain.nyc1` has rainfall information about New York City from 1869 through 1957. If we take the ratio of these two data sets, we will get valid results only for the period of time in which they overlap, namely 1890 through 1927:

```
> rain.nyc1 / corn.rain
 1890: 4.552083 2.914729 3.444444 5.356322 5.779412 2.696000
 1896: 3.084615 4.198020 4.574257 3.643564 3.648148 6.025641
 1902: 3.104938 3.936170 3.726415 3.550000 3.426087 3.220588
 1908: 3.256198 3.325000 3.516129 6.038961 4.018182 8.130435
 1914: 4.052632 2.612121 3.946237 4.212766 4.241379 5.347368
 1920: 4.586207 3.123967 5.587500 3.794393 3.000000 3.663717
 1926: 4.120690 5.394231
```

2.7 Complex Numbers

A wide variety of mathematical and engineering problems are greatly facilitated by the use of complex numbers. Of course, many users never need to use complex numbers, so feel free to skip this section if it's not of interest to you. Most S arithmetic functions correctly handle complex numbers automatically, so the S environment is a good place to work with these kinds of numbers.

A complex number consists of two parts: a real part and an imaginary part. S will read complex constants of the form $n_1 \pm n_2 i$, where both n_1 and n_2 are numeric constants. The letter i in this case has a special meaning; it represents the value of the square root of −1. For example, the following are legitimate complex numbers, suitable for reading by scan:

```
3 + 6i
4.13 + 2.7i
4.1i
6-2.4i
```

If you need to form a complex number from an expression or a variable, you can use the function complex. The real and imaginary parts of the desired number are passed through the real= and imaginary= arguments to complex. For example, to form a complex number whose real part is equal to the variable x, and whose complex part is equal to the square root of x, you could write

```
> x <- 22
> complex(real=x,imaginary=sqrt(x))
[1] 22+4.690416i
```

2.8 Libraries

Although the distributed version of S provides a wide range of functionality, there will always be specialized computations and procedures for which the necessary functions are not available. In many cases a set of functions for a particular task has been written, and may be available on your computer as an S library. A library is simply a directory containing S functions and related documentation. To see what libraries are available on your system, use the function library with no arguments; for example:

```
> library()
The following sections are available in the library:

SECTION          BRIEF DESCRIPTION

examples         Functions and objects from The New S Language
semantics        Functions from chapter 11 of The New S Language
                 . . .
```

This will provide a very brief description of the library. To find out what functions are available in a particular library, call the library function with the help=libraryname argument; for example, library(help=examples). This will display a brief description of the functions available. Finally, to allow access to the functions in the library, call the library function with the name of the library as an argument, as in library(examples). You can then use the help function in the normal way to find out more about any function or data set of interest.

2.9 Advanced Topics

2.9.1 Attributes

When working with data in S, you can usually concentrate on the values of the data themselves, without worrying about the type of data, number of elements, names of variables, and other details, because functions in S generally know how to extract that information and behave accordingly. The techniques of object-oriented programming, introduced in Section 7.6.3, extend this idea by providing different methods for different types of S objects, based on the class of an object. This type of information is stored in S objects as attributes. Each object in S minimally has the two attributes **mode** and **length**; depending on the object in question, there may be more attributes. For example, time series data, discussed in Section 2.6, have an additional attribute, known as **tsp**, which contains information about the starting time of the series and the frequency of observations. Categorical data, introduced in Section 3.11, contain an attribute called **levels**, which contains labels describing the different values that the variable can take. In addition, as described below, you can create attributes to meet your own needs, and they can be stored and accessed in the same fashion as the built-in attributes.

The function **attributes** can be used to list or change the attributes of an object. This function accepts one argument, namely the object of interest, and returns a list containing all the attributes of the object except its length and mode. (These two attributes can be accessed through the functions **length** and **mode**, respectively.) When used on the left-hand side of an assignment statement, **attributes(object)** can be set to a list with element names that correspond to the attributes to be changed. For example, the time series object **corn.yield** is part of the standard S distribution, and contains the yearly corn yield for six states over a period of 38 years. If we wish to examine its attributes, we could use the statement

```
> attributes(corn.yield)
$tsp:
[1] 1890 1927     1
```

We can see that it has the time series attribute **tsp** with the value **c(1890,1927,1)**, indicating that its values range from 1890 to 1927 with a frequency of 1 year. When we display this object, the information is displayed along with the data:

```
> corn.yield
1890: 24.5 33.7 27.9 27.5 21.7 31.9 36.8 29.9 30.2 32.0 34.0
1901: 19.4 36.0 30.2 32.4 36.4 36.9 31.5 30.5 32.3 34.9 30.1
1912: 36.9 26.8 30.5 33.3 29.7 35.0 29.9 35.2 38.3 35.2 35.5
1923: 36.7 26.8 38.0 31.7 32.6
```

The function **attr** takes two arguments: the name of the object in question, and a quoted string or character variable containing the name of the attribute in question. Like `attributes`, a call to `attr` can be used on the left-hand side of an assignment to change the value of a particular attribute of an object. We could use the `attr` function to change the value of the `tsp` attribute of `corn.yield` with the statements below; the result is reflected in the display of `corn.yield`.

```
> attr(corn.yield,"tsp") <- c(1920,1957,1)
> corn.yield
1920: 24.5 33.7 27.9 27.5 21.7 31.9 36.8 29.9 30.2 32.0 34.0
1931: 19.4 36.0 30.2 32.4 36.4 36.9 31.5 30.5 32.3 34.9 30.1
1942: 36.9 26.8 30.5 33.3 29.7 35.0 29.9 35.2 38.3 35.2 35.5
1953: 36.7 26.8 38.0 31.7 32.6
```

In addition to these two general functions, certain specialized functions can provide access to many attributes (both for query and for modification). Although all attributes can be modified as described in the last two paragraphs, the use of these specialized functions is often more convenient. The functions that provide this access to an object's attributes include `mode`, `length`, `dim`, `dimnames`, `names`, `tsp`, `class`, and `levels`. Each accepts a single argument (the object name) and will return the named attribute, or `NULL` if it does not exist. These functions can also be used on the left-hand side of an assignment statement. An additional function of interest is `structure`, which allows you to assign a value to an object and set any attributes of interest in a single function call. For example, a matrix could be created through a call to `structure` by setting its `dim` attribute:

```
> mat <- structure(c(1,4,7,9,2,8), dim=c(3,2))
> mat
     [,1] [,2]
[1,]   1   9
[2,]   4   2
[3,]   7   8
```

As an example of a user-defined attribute, recall that when a matrix is converted to a data frame, the dimnames of the matrix must be converted to valid variable names, containing no embedded blanks or special characters. Sometimes it would be useful to store the original names of variables along with the data. We can achieve this goal by creating an attribute to be stored along with the data; in this case, a good choice for a name would be `"label"`. The following code converts the data set `auto.stats` into a data frame called `auto`, and stores an attribute called `"label"` along with the data, containing the original variable names.

```
> auto <- data.frame(auto.stats)
> for(i in 1:dim(auto)[2])
+    attr(auto[,i],"label") <- dimnames(auto.stats)[[2]][i]
```

The label for an individual column can now be accessed through the `attr` function:

```
> attr(auto[,4],"label")
[1] "Repair (1977)"
```

Thus we could use the `label` attribute to create axis labels in a plot with a statement such as the following:

```
>    plot(auto[,8],auto[,10],xlab=attr(auto[,8],"label"),
+                            ylab=attr(auto[,10],"label"))
```

In this example, the `auto.stats` matrix had to be converted to a data frame before attributes could be assigned to its columns, because the individual columns of a matrix cannot have attributes of their own. The recursive nature of data frames allows each of the columns of the data frame to have its own individual attributes.

2.9.2 Access to Data

In addition to libraries, you can also access data and functions from other `.Data` directories or data frames through the use of the function `attach`. This function modifies the search list described in Section 2.4 so that you can either access S objects stored in directories other than your local `.Data` directory, or refer to variables in a data frame or list without having to explicitly state the name of the object within which the variables are to be found. For example, you may keep separate `.Data` directories for several projects, but require access to one or more of them from all the projects. Or, if you are working intensively on a single data set stored as a data frame, you may not want to be referring constantly to the frame itself, but only to the variables defined in the frame. There are two arguments to `attach`, but only the first is required. If this first argument is a character variable or quoted string containing the name of a directory, the directory is placed in the second position of your search list; the objects in the directory will then be available in the usual way, provided that there is not an object with the same name in your working database (the first directory, data frame, or list in your search list). Alternatively, a character variable or quoted string containing the full pathname of a recursive S object (such as a data frame or list) can be passed to `attach`, provided that each element of the object has a valid name. This condition will always be met for data frames. In this case, the elements of the object that is passed to `attach` will be accessible directly by their names. Finally, the (unquoted) name of a recursive object or an expression returning a recursive object, accessible through the current search list, can be passed to `attach`, making the elements of the object accessible as described above.

To remove a database from the search list, use the function `detach`. It accepts either a character variable or a quoted string containing the name of the directory or object to be removed, an integer representing the location within the search list of the directory or object to be removed, or a logical vector of the same length as the search list, with `TRUE` in those positions representing entries to be detached and `FALSE` elsewhere. An optional second argument to `detach`, `save=`, is described below.

By default, **attach** places the database in question on the search list in position 2, immediately after your working database. Thus, by default, new or modified objects will continue to be stored in the usual location, even after you have attached another directory or recursive S object. Sometimes it is necessary to attach a data directory or recursive S object as the working database, however. For example, you may be working on several projects at once, and may wish to store some objects in one directory and then switch to storing objects in some other directory. In cases like this, where the attached object is a data directory, you can simply specify the optional **pos=1** argument to **attach**. Creation of objects and modifications to existing objects will now be stored in this attached directory in the usual fashion. When you attach a recursive object, such as a data frame, as the working database using the **pos=1** argument, a local copy of the object is maintained in memory, and it is this local copy that will reflect new objects or changes. In order to retain modifications in a data frame or list attached as the working database, it must be detached and saved before the end of your S session. To do this, you should call the **detach** function, with the **save=** argument equal to a character variable or quoted string containing the name of an object in which the data frame or list should be stored. The modified data frame will then be stored in an object of that name in the second database on the search list. For example, suppose you are working with a data frame named **corn**, which contains the variables **Nitrogen**, **Phosphorus**, and **Yield**. You could attach and modify the data frame, saving the changes in a new data frame named **corn.new**, as follows:

```
> attach(corn,1)
expressions to modify Nitrogen, Phosphorus, or Yield
> detach(1,save="corn.new")
```

The object **corn.new** would now contain the modified values, and would be stored in the database that was the working database before the **corn** data frame was attached.

An additional consideration when a data frame is used as the working database results from the somewhat restrictive structure of data frames. In particular, all elements of a data frame must be of the same length. Thus, if a data frame is the working database and an object whose length is not the same as the first dimension of the data frame is modified, that object will not be saved. You can avoid this difficulty by converting the data frame into a parameterized data frame. The simplest way to do this is to assign a parameter to the data frame with the **parameters** assignment function. Suppose we wish to store a character string called **variety** along with the **corn** data frame described above. We could use the following statements:

```
> parameters(corn) <- list(Variety="")
> attach(corn,1)
> Variety <- "Supersweet"
> detach(1,save="corn.new")
```

The data frame `corn.new` will now have the variable `Variety`, equal to the specified value. Parameterized data frames are used extensively in statistical models, especially nonlinear ones. Chapter 10 presents more details on their use.

When using `attach` to modify your search list, you should be aware that S keeps an internal table of object names found in each database in the search list to provide faster access to those objects. If objects are deleted from or added to any of the databases on the search list during the current S session, this internal table is appropriately updated. However, if system commands or another S job change the contents of these databases, the tables will not be revised. You can force the revision of any member of the search list by using the function `synchronize`, with a single argument equal to the position of the database in question on the search list.

2.9.3   Direct Access to Data

The procedures outlined above all operate through modification of the internal search list, which is used by the S interpreter to resolve the names of objects to which you refer in your S statements. Occasionally, you may need to access the value of a particular object without modifying the search list. For example, if you wish to access only one or two objects from a different `.Data` directory, you may not want to include it in the search list. The function `get` can be used to access an item when you wish to explicitly state the database in which it will be found. Suppose you need access to an S object called `survey.data`, located in another user's `.Data` directory, which is named `/usr/joe/.Data`. You could store a local copy of the object in your working database with the command

```
> survey.data <- get("survey.data",where="/usr/joe/.Data")
```

assuming that you had appropriate read permissions for files in that directory.

When you need to specify which directory within your search list should be used to find an object, you can use an integer as the `where=` argument to `get`, representing the location of the directory in question in your search list. For example, the object `letters` is a character vector containing the lowercase letters of the alphabet, stored in the S system data set directory. If you had an object called `letters` in your working database, you would not be able to see the definition of the system data set `letters` by typing its name; instead, you would see the local definition. You can check on the location of the S system data set directory using the `search` command, and then use its location as the `where=` argument to `get`:

```
> search()
[1] ".Data"
[2] "/home/splus/splus/.Functions"
[3] "/home/splus/stat/.Functions"
[4] "/home/splus/s/.Functions"
[5] "/home/splus/s/.Datasets"
[6] "/home/splus/stat/.Datasets"
[7] "/home/splus/splus/.Datasets"
> letters <- c("R","x","M")
> letters
[1] "R" "x" "M"
> get("letters",where=5)
 [1] "a" "b" "c" "d" "e" "f" "g" "h" "i" "j" "k" "l" "m" "n" "o"
[16] "p" "q" "r" "s" "t" "u" "v" "w" "x" "y" "z"
```

See Section 6.11.2 for techniques to determine in which database a particular object may be found.

Another important use of **get** is to access objects through a character representation of their names, since the first argument to **get** is a character string containing the name of the object in question. For example, suppose you have sales data for a variety of products stored in data frames named **sales.jan**, **sales.feb**, and so on, and you wish to access the sales for a particular month given its abbreviation. By using the **paste** function, you can construct a character string with the appropriate name and pass it to **get** to access the actual value:

```
> month <- "feb"
> sales <- get(paste("sales",month,sep="."))
```

The data frame **sales** will then contain the contents of the data frame **sales.feb**. Note that with only one argument, **get** uses the usual search list to resolve the name you pass to it.

If you refer to an object through **get** that does not exist, an error message will be printed. To avoid this, you can use the function **exists** to check for the existence of a particular object before calling **get**. Along with the name of the object (which is required) and the optional **where=** argument, you can pass an optional **mode=** argument to determine whether an object of a particular mode exists. Possible modes include **"numeric"**, **"function"**, and **"list"**. Note that if no **where=** argument is given, **exists** will search the directories in your search list to determine whether the object in question exists.

Assignments to named objects can be made directly using the **assign** function. The first argument to **assign** is a character variable or quoted string containing the name of the object to be assigned a new value; the second argument is any valid S expression, and will be assigned to the object whose name is represented by the first argument. With no other arguments, assignments take place in the working database. To assign values to objects in other databases, use the optional **where=** argument, providing either a character value containing the name of a data directory, or an integer giving the position of the appropriate element of the search path. (Remember that you must have read/write privileges in any directories containing objects to which you are assigning values.) For example, to assign the value of a data frame called `sales` to an object called `sales.jan` in the directory `monthly/.Data`, you could use the following call to assign:

```
> assign("sales.jan",sales,where="monthly/.Data")
```

assuming you had appropriate write permissions in the named directory. Notice that the first argument is the target and the second argument is the value to be assigned, in line with the usual assignment statement (`target <- assignment`). As is the case for `get`, if you pass an integer as the `where=` argument, `assign` will carry out the assignment in the directory whose position in the search list corresponds to the argument.

When using either `assign` or `get`, remember that special operators such as square brackets (`[ ]`) and dollar signs (`$`) will be interpreted literally by `get` or `assign`; for example, the expression

```
> assign("mylist$element",20)
```

will create an object called `mylist$element`; this object will be accessible only through the `get` function, because it contains a special character. To operate on list elements through character representations of their names, you must first create a temporary copy of the list, modify this copy, and then assign it back to the original list. For example, to assign the value of 20 to the element named `element` in the list named `mylist`, you could use the following statements:

```
> tmplist <- get("mylist")
> tmplist[["element"]] <- 20
> assign("mylist",tmplist,where=1)
```

The argument `where=1` ensures that assignment takes place in the working database, even if the call to `assign` is embedded in a function.

2.9.4   Advanced Features of the scan Function

Reading Part of a File

In some cases, you may need to ignore some of the records contained in your input file.   For example, the file may contain header records that should be ignored, or you may want to read only part of the data file. The `skip=nlines` argument is an integer that tells `scan` to ignore the first `nlines` lines of the file and to start reading data after these lines. The `n=nitems` argument tells `scan` to stop reading from the file after `nitems` items have been read.

In S-PLUS, there is a `widths` argument that allows you to specify the width of fixed-format data fields in your input file. You can pass a vector of values through this argument, using one element for each element in the `what=` argument to specify how many columns from your input file should be read for each variable.

Reading Data Directly into a List

By using the **what=** argument of the **scan** function, you can read data from a file or from the keyboard directly into a list. For example, suppose we had a data file named **scores** containing three values per line, the first being a name and the second and third being test scores. Sample lines from the file might look like these:

```
joe 20 49
fred 18 32
harry 19 42
sue 25 51
```

We could create a list called **grades** containing three elements, **name**, **score1**, and **score2**, with the following call to **scan**:

```
> grades <- scan("scores",what=list(name="",score1=0,score2=0))
```

If a particular field in the input file is to be skipped, the object **NULL** can be used as a component of the **what=** list. When scan is used with a **what=** argument, the concept of a line within a file takes on additional meaning. Recall that, by default, **scan** reads your file as a continuous stream, with data values separated by a known separator, and newlines having no special meaning. When you read multiple items, as in the example above, **scan** expects all of the values specified in the list to be on the line. There is no problem with having multiple sets of values on the same line, but if you break up a set of values across lines, you will see an error message similar to the following:

```
Error in scan(, what = list(names = "", score1 =: Expected 3
fields on line 1, found 2
```

This behavior can actually be useful, because data files are usually organized with records residing on the same line, so an unsuccessful call to **scan** may be an indication that the file is not organized the way it should be, or that the call to **scan** was incorrect. If you don't want this kind of checking to take place, and wish to allow records to span multiple lines or to ignore some fields on each line, two additional options to **scan** provide some flexibility in the way it extracts data from the individual lines in your file. The **multi.line=T** option lifts the restriction that records must reside on a single line. (Notice that if you use the **multi.line=T** option when reading data from the keyboard, you must terminate your input with a Ctrl-D character (press Control and D at the same time), instead of the usual blank line. When you are reading from a file, no changes are required.) The **flush=T** option tells **scan** to stop reading from a line after all the items in the **what=** argument have been exhausted, allowing, for example, comments at the end of each data line in the file.

Exercises

1. Using the **rbinom** function, generate a sample of size 10 from a binomial distribution representing 20 coin flips with a probability of 0.4. Use the **stem** or **hist** function to plot the distribution of values. Repeat with different sample sizes, ranging up to 100 or so. It is often said that the normal distribution is a good approximation to the binomial. From the distributions you produced, at what value of the sample size does the distribution appear to be normal? Repeat the experiment using a probability of 0.05. Try the same exercise with other distributions, such as the lognormal (**rlnorm**).

2. Execute the **library()** command under your version of S or S-PLUS to see if any libraries are installed on your system. If you find any of interest, examine the **help** files and try some of the functions. Is there any way to distinguish these functions from those that are an actual part of the S language?

3 Operators and Functions

3.1 Arithmetic Operators

Table 3.1 displays the arithmetic operators available in S. In each case (except for the unary minus), the operator expects two operands, one on the left side of the symbol and one on the right. Operands can be combined to create complex expressions.

Table 3.1 Arithmetic Operators

Operator	Function	Operator	Function
+	addition	^	exponentiation
−	subtraction	%/%	integer divide
*	multiplication	%%	modulus
/	division	:	sequence
%*%	matrix multiplication	−	unary minus

When both operands are scalars, the result of the operation is a scalar. But the arithmetic operators are also designed to work intelligently with matrices and vectors, provided that the modes and lengths of the arguments are compatible. For example, all the columns of a matrix must contain the same number of elements, so when any of the operators in Table 3.1 (except for matrix multiplication) are applied to the columns of a matrix, the result will be a vector of the same length as the two columns. Using the matrix `auto.stats`, column 3 contains the repair rating (on a scale of 1 to 5) for the cars in 1978, and column 4 contains the same rating in 1977. Thus we can create a vector of the differences between these ratings with the following statement:

```
> rating.diff <- auto.stats[,3] - auto.stats[,4]
```

The length of the vector `rating.diff` will be the same as the number of rows in the `auto.stats` matrix, because that is the length of a column in that matrix. When the arguments to the operators are a scalar and a vector, the result

43

will be of the same length as the vector, each element containing the result of applying the operator to the scalar and the corresponding element of the vector. To convert the turning-circle measurement of column 10 of the `auto.stats` matrix from inches to centimeters, we can multiply the vector containing the measurements by the scalar conversion factor, namely 2.54 centimeters per inch:

```
> turning.cm <- 2.54 * auto.stats[,10]
```

The result, `turning.cm`, is a vector of the same length as `auto.stats[,10]`. Note that we could equivalently have referred to the turning circle as `auto.stats[,"Turning Circle"]`, since `"Turning Circle"` is the dimname corresponding to column 10 of the `auto.stats` matrix.

Care should be taken when using the operators on vectors of different lengths, because S will recycle the elements of the shorter vector to provide a result that is the same length as the longer vector. No message will be printed, unless the elements of the smaller vector cannot all be used the same number of times, as the following examples demonstrate.

```
> x <- 1:10
> y <- 1:5
> z <- 1:7
> x - y
[1] 0 0 0 0 0 5 5 5 5 5
> y * x
[1]  1   4   9  16  25   6  14  24  36  50
> z %% x
[1] 0 0 0 0 0 0 0 1 2 3
Warning messages:
  Length of longer object is not a multiple of the length
  of the shorter object in: z %% x
```

When subtracting `y` from `x`, the elements of `x` are silently used twice, since the length of `y` is exactly twice the length of `x`; the same silent reuse takes place in multiplication. When the length of the longer operand is not an even multiple of the shorter one (such as `z %% x`), elements are still reused, but the warning message is printed.

Similar element-by-element operations will be performed when the operands are both matrices or when one operand is a matrix and the other a scalar. In the case of two matrices (or other arrays), the dimensions of the matrices must be identical.

Notice that the multiplication operator (`*`) will perform element-by-element multiplication of two matrices provided that they have the same number of rows and the same number of columns. The usual function of matrix multiplication is obtained through the matrix multiplication operator (`%*%`). In order for two matrices to be conformable for matrix multiplication, the number of columns in the first matrix must be the same as the number of rows in the second matrix. The resultant matrix will have the same number of rows as the first matrix and the same number of columns as the second matrix. If the matrix A is of dimension $n \times m$, and the matrix B is of dimension $m \times p$, then the i, jth element of the product of the two matrices is defined by the following formula:

$$C_{ij} = \sum_{k=1}^{m} A_{ik} B_{kj} \quad \text{for } i = 1, ..., n, \quad j = 1, ..., p$$

The integer divide operator (`%/%`) performs division between its two operands, truncating its result in the process. If you are interested in rounding instead of truncating, you can use the functions **round** or **signif**. Related functions include **floor** and **ceiling**, which return the closest integer not greater than or not less than their argument, respectively. (See Section 3.6.4 for more information about these functions.)

The modulus operator (`%%`) returns the remainder that would result if its first operand were divided by its second operand. If the second operand is zero, the result is simply the first operand. Because of the limited precision inherent in storing noninteger numbers, you should use the modulus operator on nonintegers with caution.

The unary minus changes the sign of its operand. It can be applied either to numbers or to valid S expressions. Do not confuse the unary minus operator with the logical "not" operator, discussed in the next section.

The sequence operator is discussed in Section 2.3.

3.2 Logical Operators

In addition to the arithmetic operators mentioned above, S provides a full range of logical operators. The logical operators are displayed in Table 3.2.

Table 3.2 Comparison and Logical Operators

Comparison Operator	Function	Logical Operator	Function
>	greater than	&	elementwise and
<	less than	|	elementwise or
<=	less than or equal to	&&	control and
==	equality	||	control or
!=	non-equality	!	unary not

S provides the basic logical comparison operators for equality and inequality, as shown in the first column of Table 3.2. These operators will evaluate the relationship between their operands and return TRUE, FALSE, or NA as appropriate. Like the arithmetic operators, these operators are smart about dealing with scalars, vectors, and matrices. It should be kept in mind that most operations, including comparisons, involving a missing value (NA) will result in a returned value of **NA**. In particular, regardless of the value of **x**, the statement **x == NA** will always result in a value of **NA**. When it is necessary to test for a missing value, the function **is.na** should be used instead of the usual equality operator, **==**.

As can be seen in the second column of Table 3.2, there are two different logical operators for the functions of "and" and "or." The first set of operators, labeled "elementwise" in the table, operate in much the same way as the arithmetic operators described above. When both operands are scalars, the result is a scalar. When both operands are vectors, the result is a vector of length equal to the larger of the two operands, where each element is the result of applying the operator to the corresponding elements in the operands. When the operands are a vector or matrix and a scalar, the result is an object like the vector or matrix, each element of which represents the result of the operation using the element of the original vector or matrix and the value of the scalar. Finally, for two matrices, the result is a matrix whose elements represent the result of applying the operator to the corresponding elements of the original matrices, provided they have the same dimensions. In addition to the listed operators, the function **xor** accepts two logical arguments and returns the result of an exclusive or; that is, **xor(e1,e2)** returns **TRUE** if and only if exactly one of **e1** and **e2** is true. Its behavior with vectors and matrices is identical to that of the other elementwise operators.

The second set of operators, labeled "control" in the table, expect an object consisting of a single logical value as their first operand. If this value is **TRUE**, the "control and" operator (**&&**) evaluates its second operand and returns its value, coercing to a logical value if necessary. If the first operand is **FALSE**, however, the **&&** operator returns **FALSE** without evaluating its second operand. In a similar fashion, the "control or" (**||**) operator evaluates its first operand and, if it is **FALSE**, the operator evaluates and returns its second operand, coercing to a logical value if necessary. If the first operand is **TRUE**, the operator returns **TRUE** without evaluating its second argument. Regardless of the lengths of the operands, the control operators return a single value of **TRUE**, **FALSE**, or **NA**. If the first operand contains more than one element, or if the second argument is evaluated and found to contain more than one element, only the first element is used in either case. The "control and" (**&&**) operator is useful in avoiding unnecessary or illegal operations, because its second operand will not be evaluated unless its first operand is **TRUE**. Suppose we wish to test whether the sum of the elements of an arbitrary vector **x** is equal to zero. We could avoid calling the **sum** function with a nonnumeric vector by using a statement such as the following:

```
> is.numeric(x) && sum(x) == 0
```

If **x** were a character vector, since the first condition would be **FALSE**, the expression would return **FALSE** without actually calling the **sum** function.

Two functions that are very useful when combining logical objects are **any** and **all**. These functions take an atomic logical object of any size and return a single logical value of **TRUE** if any or all of the elements are **TRUE**, respectively.

The returned value from these functions is suitable for use either with the control (&& and ||) operators or with the elementwise (& and |) operators.

For example, suppose we wish to determine whether any of the countries in the `saving.x` data set have populations of which more than 50% are under 15 years of age. The first column in the data set contains this percentage, so we could use the S statement

```
> any(saving.x[,1] > 50)
[1] F
```

In a similar fashion, we could test whether any of these nations have disposable income greater than $1000 and growth greater than 9 percent with the statement

```
> any(saving.x[,"Disp. Inc."] > 1000 & saving.x[,"Growth"] > 9)
[1] F
```

Notice the difference between the preceding statement, which first logically compares two vectors of length 50 before applying the **any** function, and the following statement, which applies **any** to two vectors of length 50 before combining them with the and (**&**) operator:

```
> any(saving.x[,"Disp. Inc."] > 1000) & any(saving.x[,"Growth"] >9)
[1] T
```

The "unary not" operator (!) changes the sense of its argument—that is, TRUE becomes FALSE and FALSE becomes TRUE—when the operator is applied to a single operand.

3.3   Special Operators

S also offers some operators that have special functions discussed elsewhere in this book. The assignment operator <- has been discussed in Section 1.4. The special assignment operator <<-, which forces assignment to a data directory, is useful in writing functions, and is covered in Section 6.9. The tilde (~), which is used to define statistical models, is discussed in Chapter 10.

3.4   Operator Precedence

Because of the wide range of operators and expressions available within S, it is conceivable that certain combinations of operators and expressions might be interpretable in more than one way. For example, if we write an expression such as

```
> - 3 * 5 / 4 + 2
```

there are a number of ways this might be interpreted. One possibility would be to multiply -3 by 5, add 4 and 2, then divide. Another might be to divide 5 by 4, multiply by 3, then add 2 and change the sign of the result. To ensure that there is one predictable way in which complex expressions will be evaluated, rules for the ordering of various operations are generally established. These rules, known as precedence rules, allow you to determine the order in which operations will be carried out in a complex expression. For example, in the expression above, first the unary minus will be applied to 3, then -3 will be multiplied by 5, followed by the division of this result by 4, and finally the addition of 2, to yield the final answer. However, the most important thing to know about precedence rules is that you never need to memorize them or even look them up, because you can freely use parentheses (()) to make sure that S understands exactly what you are doing. For example, we could rewrite the expression above as

```
> - 3 * (5 / 4) + 2
```

to make absolutely clear the order in which we want the operations to be performed.

Even so, it is useful to know how S will interpret your statements when they do not include parentheses inserted for clarity. For example, is the expression 1:n+1 interpreted as a sequence running from 1 to n+1, or a sequence running from 1 to n, with 1 added to each element? In Table 3.3, operators near the top of the table will be applied to their operands before the operators below them in the table. If two operations in an expression have the same precedence, they are evaluated from left to right. Thus, in the example above, the sequence operator (:) will create the sequence from 1 to n, and then 1 will be added to each element. Similarly, the expression 1:10[4] will not return the value 4; an error message will be displayed, because the subscript operator ([ ]) binds to its argument more tightly than the sequence operator (:); thus S interprets the expression to mean a sequence that runs from 1 to the fourth element of 10, which of course doesn't exist. The proper way to extract the fourth element of the sequence is (1:10)[4], ensuring that the sequence operator is executed before subscripting takes place.

Table 3.3 Precedence Rules for S Operators

Operator	Name
$	List and data frame extraction
[[[	Subscripts
^	Exponentiation
-	Unary minus
:	Sequence
%*% %/% %%	Matrix multiplication, integer division, modulus
* /	Multiplication and division
+ -	Addition and subtraction
< > <= >= == !=	Logical comparisons
!	Unary not
& | && ||	Logical and, control and, and, or
<- <<- _	Assignment

3.5 Introduction to Functions

Most of the capabilities of the S language are achieved through the use of functions. The remaining sections of this chapter provide brief descriptions of many of these functions. You can skim these sections to get a feel for what functions are available in S, and come back to them later for more details when you need the solution to a particular problem. The remainder of this section describes the basic information you need in order to call functions, and a brief introduction to how to write your own functions to combine, modify, or add to the collection of functions already available.

Arguments to functions are enclosed in parentheses and separated by commas. Even if no arguments are being passed to a function, the parentheses are required. For example, to call the function max, which returns the maximum value of its arguments, with the three arguments x, y, and z, you would use the expression

```
> max(x, y, z)
```

The spaces between the commas and the arguments are optional. However, to call the function objects with no arguments, you would type

```
> objects()
```

Each of the arguments to a function has a name, so you can specify arguments in any order and make sure that the function will understand what you want. To specify a named argument to a function, use an equal sign between the name of the argument and the value. (These equal signs do not permanently assign values to a variable; they simply associate a name with an argument for the duration of the function call.) To get an equally spaced sequence of length 11 ranging from 0 to 100, we could call the seq function as follows:

```
> pts <- seq(0,100,length = 11)
> pts
 [1]   0  10  20  30  40  50  60  70  80  90 100
```

Spaces between the equal signs and the values are optional. It was necessary to explicitly use the name length, because the seq function was defined to have its arguments in a different order. You can check the online help files to see the default order of arguments for any function. Continuing with the seq example, the online help file shows the definition of this function as

```
        seq(from=, to=, by=, length=, along=)  # as function
```

This means that if we passed 11 as the third argument without a name, the function would interpret it as being the by argument.

Not all the arguments to functions are required; many of the functions are written to provide reasonable defaults for arguments that are not specified. In the help files, these defaults are shown as values to the right of the equal sign. For example, the function rnorm, which generates random samples from a normal distribution, is defined in the help file as

```
        rnorm(n, mean=0, sd=1)
```

This means that if no **mean** argument is given, it will default to a value of **0**; similarly, if no **sd** value is given, a default value of **1** will be used. If we call **rnorm** with three unnamed arguments, it will assume that they represent **n**, **mean**, and **sd**, in that order. Thus the statement

```
> norm.sample <- rnorm(20,10,5)
```

will produce a vector of length 20, containing a random sample from a normal distribution with mean 10 and sd (standard deviation) 5, since that is the order in which the arguments were specified in the function definition. It is exactly equivalent to the following call:

```
> norm.sample <- rnorm(n=20,mean=10,sd=5)
```

The complete definition of any function, including default values for its arguments, can be displayed by typing the name of the function.

One of the most useful features of the S language is the ability to extend the language by writing new functions. These functions behave exactly the same as the built-in functions described above, so once you've decided on a way to attack a problem in S, writing a function is just a natural extension of the steps you would take interactively. Functions are discussed in much greater detail in Chapter 6, but it will be helpful to mention some basic information about them here. Although the construction of more complex functions will be deferred to Chapter 6, even very simple functions, consisting of a single line, can be very valuable. For example, suppose you would like to create a function that raises a number to the third power, and you want to call this function **cube**. The following one-line definition will do the job:

```
> cube <- function(x)x^3
```

The argument **x** is known as a dummy argument; it can take any value, and its existence is restricted to the environment of the function call. Accordingly, you can use any name for a dummy argument without fear of overwriting objects you're storing in your data directory.

The sections that follow present some of the basic functions available in S, grouped by category. Since S is a constantly evolving program, such a presentation can never be complete. For this reason, the online help facility, discussed in Section 1.2, should stand as the final authority regarding available functions. For each of the functions discussed below, additional information is always available through the online help facility.

3.6 Mathematical Functions

3.6.1 General Mathematical Functions

S provides a set of basic mathematical functions, as summarized in Table 3.4. They all accept vector arguments, and will return a missing value (**NA**) if they are passed an argument outside the acceptable range of the function.

Table 3.4 Mathematical Functions

Function	Description
abs	Absolute value
exp	Exponential (e to a power)
gamma	Gamma function
lgamma	Log of gamma function
log	Logarithm
log10	Logarithm (base 10)
sign	Signum function
sqrt	Square root

The function `log` is implemented with an optional argument called `base`, which defaults to the value `e` (2.71828...), and can be used to obtain logarithms for any base. The signum function, `sign`, returns −1, 0, or 1, depending on whether its argument is less than, equal to, or greater than zero, respectively.

3.6.2 Trigonometric Functions

Table 3.5 shows the trigonometric functions available in S. All of these functions will accept vector or array arguments, and return objects of the same mode and size with the function applied to each element. When any of the functions are passed an out-of-range value, they will return a missing value (`NA`). Note that each function beginning with the letter `a` is the inverse of the same function without the `a`.

Table 3.5 Trigonometric Functions

Name	Function	Name	Function
cosine	cos	arc cosine	acos
sine	sin	arc sine	asin
tangent	tan	arc tangent	atan
hyperbolic cosine	cosh	arc hyperbolic cosine	acosh
hyperbolic sine	sinh	arc hyperbolic sine	asinh
hyperbolic tangent	tanh	arc hyperbolic tangent	atanh

3.6.3 Functions for Complex Numbers

Five functions are available in S to assist in manipulating complex numbers in S. Each is capable of handling vector- or array-valued arguments. In the descriptions that follow, it is assumed that the argument to each of the functions is a complex number of the form $a + bi$. The function `Arg` accepts a complex value and returns the argument of that value—that is, the arc tangent of b/a, multiplied by the sign of b. This quantity is also known as the phase angle when the complex number is viewed in polar coordinates. The function `Conj` returns the complex conjugate of a complex number—that is, the value $a - bi$. The function `Mod` returns the modulus of a complex number—that is, $\sqrt{a^2 + b^2}$. Finally, the functions `Re` and `Im` return the real (a) and imaginary (b) portions of a complex number, respectively.

3.6.4 Functions for Rounding and Truncating

Five functions are provided in S for the purposes of rounding and truncating numbers. All of them operate on scalars, vectors, or matrices, returning a like object containing the rounded or truncated values. In addition, when passed a missing value (`NA`), each will return `NA`. The difference between rounding and truncating is that in rounding, the magnitude of the rounded value may be smaller or larger than that of the original value, depending on whether or not the last digit to be rounded is greater than or equal to 5, whereas the magnitude of a truncated value is always smaller than or equal to that of the original value. The function `round` accepts an argument called `digits`, in addition to the data that are to be rounded, which specifies the number of digits after the decimal point in the rounded result. Thus, with the default value of 0, it will round to integer values, whereas with a value of, say, 2, it will round to the nearest hundredth. Negative values may also be passed as `digits` arguments to `round`, giving rounded values to the nearest 10 (`digits=-1`), 100 (`digits=-2`), and so on. A related function is `signif`, which performs a function similar to that of `round`, the primary difference being that the `digits` argument to `signif` is interpreted as the number of significant digits without regard to the placement of the decimal point.

The function `ceiling` returns the smallest integer greater than or equal to its argument; `floor` returns the largest integer less than or equal to its argument. Because this creates an apparent inconsistency when dealing with negative numbers (for example, `floor(1.5)` is 1, but `floor(-1.5)` is -2), the function `trunc` can be used to consistently truncate numbers to the nearest integer in absolute value. In other words, `trunc` behaves like `floor` for positive values and like `ceiling` for negative values.

3.6.5 Mathematical Functions for Matrices

Because one of the strengths of the S language is its ability to operate on arrays of values with single commands, it is not surprising that a number of functions are provided to perform most common matrix manipulations. Table 3.6 lists some of these functions.

Table 3.6 Matrix Functions (Linear Algebra)

Function	Description
chol	Cholesky decomposition (sometimes called root)
crossprod	Matrix crossproduct
diag	Create diagonal matrix or extract diagonal values
eigen	Eigenanalysis (sometimes called latent roots)
outer	Outer product of two vectors
scale	Scale the columns of a matrix
solve	Solve system of linear equations; find inverse
svd	Singular value decomposition
qr	QR orthogonalization
t	Transpose

The diag function performs a number of tasks, depending on whether its single argument is a scalar, a vector, or a matrix. When passed a scalar, diag returns an identity matrix of the given dimension. (An identity matrix is a square matrix whose diagonal elements are all 1 and whose off-diagonal elements are all 0.) When the argument to diag is a vector, it returns a diagonal matrix whose nonzero elements are those of the vector. Given a matrix, diag returns a vector containing the diagonal elements of the matrix. The length of the vector returned in this case is the minimum of the number of rows and the number of columns in the input matrix.

The function chol performs a Cholesky decomposition of a square, symmetric, positive definite matrix A, finding an upper triangular matrix U such that

$$A = U'U$$

where the prime symbol ($'$) indicates the transpose of the matrix it follows. One important use of the Cholesky decomposition is in generating correlated random samples. Suppose we wish to produce a random sample with variance A. If we generate a vector x of length k containing uncorrelated random variables, and multiply x by a matrix U, the variance of the resulting vector will be $U'U$. Thus, if we multiply the vector x by the Cholesky root of A, the resulting vector will have the desired variance. Notice that the preceding discussion dealt with a single vector of length k. If n correlated multivariate normal random samples are desired, then an $n \times k$ matrix X can be post-multiplied by the transpose of U, as in the following example:

```
> x <- matrix(rnorm(15),5,3)   # 5 uncorrelated 3-vectors
> v <- matrix(c(1,.7,.4,.7,1,.7,.4,.7,1),ncol=3,byrow=T)
> vc <- chol(v)
> tx <- x %*% t(vc)   # 5 correlated 3-vectors
```

The function eigen operates on a square matrix and returns a list containing two components: values, which contains the eigenvalues, and vectors, which contains the eigenvectors. If the matrix passed to **eigen** is not symmetric, the eigenvalues and eigenvectors may be complex.

Closely related to **eigen** is the function **svd**, which performs a singular value decomposition. For an arbitrary matrix X, the singular value decomposition is defined as

$$X = UDV'$$

where U and V are orthogonal matrices and D is a diagonal matrix with the singular values along the diagonal. The singular values are the square roots of the eigenvalues of the matrix $X'X$. The **svd** function returns these quantities in a list whose elements are named **u**, **v**, and **d**.

The function **qr** performs a QR orthogonalization on an arbitrary matrix X, defined as

$$X = QR$$

where Q is an orthogonal matrix ($Q'Q = I$, the identity matrix) and R is a upper triangular matrix. The results are returned in a list containing two elements: **qr** and **qraux**. Although these values are usually not suitable for direct use, the function **qr.qy** is provided to give the product of Q and another matrix. Its first argument is the object returned from the call to **qr**; its second argument is the matrix to be multiplied by the orthogonal matrix of the decomposition. To get the Q and R matrices as described above in the definition of the decomposition of a matrix X, the following S statements can be used:

```
> x.qr <- qr(x)
> q <- qr.qy(x.qr,diag(1,nrow(x)))[,1:ncol(x)]
> r <- x.qr$qr[1:ncol(x),]
> r[row(r) > col(r)] <- 0
```

Another important use of **qr** is to calculate the determinant of a square matrix. Although no S function is provided to calculate the determinant, you can use the following function, based on the QR decomposition, provided the matrix **x** is square:

```
> det <- function(x)
+    prod(diag(qr(x)$qr)) * ifelse(nrow(x) %% 2,1,-1)
```

In addition to the matrix multiplication operator described in Section 3.1, S provides two other functions for matrix products. The function **crossprod** takes one or two arguments. Given a single vector or matrix argument X, **crossprod** returns $X'X$. Given two arguments X and Y, **crossprod** returns $X'Y$. Note that the inner product (sometimes called the dot product) of two vectors x and y, both of length n, defined as $\sum_{i=1}^{n} x[i]*y[i]$, can be obtained by calling **crossprod** with arguments x and y.

The function **outer** produces the outer product of two vectors. Given a vector x, of length n, and a vector y of length p, **outer** returns an $n \times p$ matrix whose i, jth element is $x[i] * y[j]$. The **outer** function can be generalized to perform more complex functions than multiplication; see Section 7.1 for more details.

A third commonly encountered matrix product is the Kronecker or direct product. In S-PLUS, the built-in function **kronecker** will calculate such a product from two matrices **a** and **b**. If the function is not available, the following function can be used to calculate a direct product:

```
> kron <- function(a,b)
+   matrix(aperm(outer(a,b,"*"),c(3,1,4,2)),nrow(a)*nrow(b))
```

Systems of linear equations can be solved using the function **solve**. To solve a system of equations $Ax = b$, where A is a $p \times p$ matrix of known coefficients and b is an n-vector of known right-hand values, for the unknown p-vector x, call **solve** with the two arguments A and b. When called with a single matrix argument, **solve** will return the inverse of the matrix, provided that it is square and the inverse exists.

The function **scale** standardizes the columns of a matrix. By default, calling **scale** with a single argument representing a matrix will return a similarly dimensioned matrix with the column mean subtracted from each element, and each element divided by the column standard deviation. This results in a matrix for which the mean of each column is 0 and the standard deviation of each column is 1. Two optional arguments, **center** and **scale**, give further control over the scaling. If **center** is FALSE, then no subtraction is performed; if **scale** is FALSE, then no division is carried out. Alternatively, either argument can be a vector of length equal to the number of columns in the matrix to allow subtraction and/or division by values other than the defaults.

3.7 Functions for Sorting

The S function **sort** accepts a vector as input and returns a vector containing the sorted values. Missing values (**NA**s) are generally removed before sorting and are not replaced. In S-PLUS, however, there is an additional argument to **sort**, named **na.last**, that allows missing values to be handled in one of three ways. If **na.last** is TRUE, then missing values are placed at the end of the sorted vector; if **na.last** is FALSE, they are placed at the beginning; and if **na.last** is set equal to **NA**, the missing values are discarded. Numeric values of infinity (**Inf**), both positive and negative, sort in their appropriate places. When passed character data, **sort** uses the standard ASCII collating sequence.[1] Note that **sort** operates correctly only on vectors; if you pass **sort** an array, it

---

[1]In the ASCII collating sequence, digits are followed by uppercase letters, then lowercase letters. Special characters are scattered between the digits and the letters in a somewhat unpredictable way. On most UNIX systems, the command man ascii will produce a table showing all the details.

will return a vector containing the sorted values of the array without regard to their placement in the original array. Because the **sort** function always sorts its arguments in ascending order, a sorted vector can be passed to the **rev** function if descending order is desired.

Perhaps a more common task than actually sorting a vector is rearranging one object to correspond to the sorted order of one or more other vectors. For example, when a line is drawn on a graph, the x values must be given in ascending order, with the y values reordered accordingly. This functionality is achieved in two steps. First a permutation vector of the indices corresponding to sorted order is created, and then that vector is used as a subscript for the object to be rearranged. The permutation vector is produced by the **order** function. This function accepts a variable number of vector arguments and returns a vector of indices, which, if used as a subscript of the vectors in question, will place the elements in ascending order. The ordering is based on the first vector passed to **order**; any additional vectors are used to order observations that have equal values in the first vector. Perhaps the **order** function can best be understood by noting that x[order(x)] is exactly the same as sort(x).

As an example, suppose we have a matrix of numbers and we wish to sort the rows based on the values in the first column of the matrix. Note that it is not possible to use the **sort** function directly, since it will treat the matrix as a vector. Suppose the matrix in question is as shown below:

```
> x
       [,1] [,2] [,3]
 [1,]    4    8   12
 [2,]    5    7    7
 [3,]    6    3   19
 [4,]    2    4    8
 [5,]    4    9    4
 [6,]    1    6    5
 [7,]    7    1    2
 [8,]    3    3   10
```

Whereas it is usually not necessary to explicitly save the result of the call to **order**, we will examine it here to understand how the matrix will be rearranged. Since we wish to reorder the matrix based on the values in the first column, we produce the permutation vector for that column through a call to **order**:

```
> oo <- order(x[,1])
> oo
[1] 6 4 8 1 5 2 3 7
```

We see that the sixth element is the smallest, followed by the fourth, the eighth, and so on. We can now use this vector as the first subscript of **x** to rearrange the matrix in the appropriate order. (See Section 5.1.2 for information on matrix subscripts.)

```
> x[oo,]
      [,1] [,2] [,3]
[1,]    1    6    5
[2,]    2    4    8
[3,]    3    3   10
[4,]    4    8   12
[5,]    4    9    4
[6,]    5    7    7
[7,]    6    3   19
[8,]    7    1    2
```

For this example, the same effect could be achieved more compactly through the statement x[order(x[,1]),].

A closely related task to ordering is ranking. The function **rank** accepts a vector of values and returns a vector of ranks; that is, the value in the output vector corresponding to the smallest value in the input vector is 1, the next smallest is 2, and so on. In the case of a tie, an average rank is returned. This is illustrated in the following example:

```
> xx <- c(4,5,2,7,3,5,8)
> rank(xx)
[1] 3.0 4.5 1.0 6.0 2.0 4.5 7.0
```

Note that the two values of 5 (in positions 2 and 6) would have been the fourth and fifth ranks, respectively. Since they are equal, each of them is assigned the average rank of **4.5**. In S-PLUS, the optional argument **na.last** can be used in the same fashion as with the **sort** function.

3.8 Functions for Data Manipulation

Much of the data manipulation required for routine analysis can be achieved through techniques outlined in this and other chapters, but there are a few functions that are especially useful for certain data management and manipulation tasks. These are summarized in Table 3.7.

The function **append** performs a function similar to that of the combine function, **c**, but offers additional flexibility. Unlike **c**, **append** will combine only two objects at a time, but it offers an optional third argument, named **after**, which allows you to specify at what position in the first argument the values in the second argument should be appended. (By default, **after** is set to the length of the first argument, so it will perform concatenation in the usual way.) If **after** is less than or equal to zero, the second argument is placed before the first argument. As a simple example, consider two vectors, consisting of the integers from 1 to 10 and from 101 to 105. The following examples show the use of **append**, with and without an **after** argument:

Table 3.7 Data Management and Manipulation Functions

Function	Description
append	Combine vectors
duplicated	Extract duplicated values
match	Match values in pairs of vectors
pmatch	Partial matching of values
replace	Replace values in vectors
unique	Extract unique values

```
> x <- 1:10
> y <- 101:105
> append(x,y)
 [1]   1   2   3   4   5   6   7   8   9  10 101 102 103 104 105
> append(x,y,after=5)
 [1]   1   2   3   4   5 101 102 103 104 105   6   7   8   9  10
> append(x,y,after=-1)
 [1] 101 102 103 104 105   1   2   3   4   5   6   7   8   9  10
```

The function **replace** requires three arguments. The first is a vector that will have some of its values replaced; the second is a vector of indices into the first argument where values will be replaced; and the third is a vector of values to be used in the replacement. We could replace the first, third, and fifth values of **x** with zeroes using the following statements:

```
> x <- replace(x,c(1,3,5),0)
> x
 [1]  0  2  0  4  0  6  7  8  9 10
```

The somewhat more complex case of data merging based on a common variable can be achieved by using the function **match** to determine which observations in one matrix or data frame have common values with a variable in another matrix or data frame. Subsetting expressions can then be used to rearrange the data so that they can be combined based on common values of the variable. The first argument to **match** is a vector containing the values to be "looked up" in the vector that is presented as the second argument; the function returns a vector of the same length as the first argument, containing the indices in which matches occurred relative to the second argument. If a value in the first argument is not matched anywhere in the second argument, by default a missing value (**NA**) is returned in that position. The optional **nomatch** argument to the **match** function can be used to return a value other than missing when no match is found.

To illustrate, consider a data frame containing state names and abbreviations, formed by combining the two data sets **state.name** and **state.abb**. The goal is to add a new variable to a data frame formed from the **state.x77** data frame, which contains the appropriate abbreviations. In this example, all data sets are sorted in the same order, so the matching is not strictly necessary, but the same techniques would apply even if the two sets were ordered differently. First we create the two data frames:

```
> ss <- cbind(state.name,state.abb)
> state <- data.frame(state.x77)
```

Next, we find the indices in the row names of **state** whose values match those in the first column of **ss**, which also contain the full names of the states. These indices can be used to select the abbreviations from the second column of **ss** that correspond to the values of the **state** data frame, to create the new variable **state$Abb**. The argument **nomatch=0** is used so that values found in the data to be merged, but not in the target data, will not produce any assignments. (Recall that subscripts of 0 are ignored, whereas subscripts equal to **NA** will produce an object all of whose values are missing. See Section 5.1.1 for more information.)

```
> locs <- match(ss[,1],row.names(state),nomatch=0)
> state$Abb <- ss[locs,2]
> cbind(dimnames(state)[[1]],state$Abb)
              [,1]          [,2]
[1,] "Alabama"      "AL"
[2,] "Alaska"       "AK"
[3,] "Arizona"      "AZ"
[4,] "Arkansas"     "AR"
[5,] "California"   "CA"
[6,] "Colorado"     "CO"
```
                       . . .

The function **pmatch** is used in exactly the same way as **match**, but does not require an exact match. If values in the first argument to **pmatch** have been truncated, but a truncated value uniquely matches the initial substring of one of the entries in the second vector, **pmatch** will return the appropriate location. Continuing with the example of state names, suppose that some names had been abbreviated in the vector **anames**:

```
> anames <- c("Cal","Flo","Mai","Ala","Col","P")
> pmatch(anames,dimnames(state)[[1]])
[1]   5  9 19 NA  6 38
```

Note that **pmatch** could not find a match for the string **"Ala"**, since two state names begin with those three letters. However, it could successfully match the string **"P"** because only one state begins with the letter *P*.

Two functions are provided to deal with the case of multiple occurrences of values in vectors. The function **duplicated** accepts a single argument and returns a logical vector containing **TRUE** in those positions that contain a value that appeared previously in its argument, and **FALSE** elsewhere. The returned object is the same length as its argument. The function **unique** also accepts a single argument, and returns a vector containing all the values found in its arguments, with each value appearing exactly once. Thus, the length of its output is equal to the number of unique values in the vector (and can be accessed by calling the **length** function with the output from **unique**). For example, to find the unique values of engine displacement in the `auto.stats` data set, we could use the statements given below. (The `drop=F` argument suppresses the printing of the dimnames corresponding to the unique values, and does not affect the function being called.) The number of unique values is stored in the variable `nvals`.

```
> dvals <- unique(auto.stats[,"Displacement",drop=F])
> dvals
  [1] 121 258 131  97 196 350 231 111 425  98 250 200 151 119  85
 [16] 146 318 225 105 140 107  91 400 302  86 163 156  79 134  89
 [31]  90
> nvals <- length(dvals)
> nvals
  [1] 31
```

Notice that the output from **unique** is in the same order as that in which the values were encountered in the input vector. To produce a vector containing the sorted unique values, the function **sort** can be called with the output from **unique**.

3.9 Functions for Simple Statistics

Although the topic of statistical modeling will be covered in detail in Chapter 10, there are some simple functions in S that provide basic descriptive statistical quantities, summarized in Table 3.8. In addition, a variety of functions for multivariate statistical techniques are described in Section 10.13.

Table 3.8 Statistical Functions

Function	Description	Comments
cor	Correlation	One or two arguments plus optional **trim=**
cumsum	Cumulative sum of a vector	
mean	Arithmetic mean	Optional **trim=**
median	Median	Actually the 0.50 quantile
min	Smallest value	Accepts variable number of arguments
max	Largest value	Accepts variable number of arguments
prod	Product of elements of a vector	Accepts variable number of arguments
quantile	Quantiles	Similar to percentiles
range	Minimum and maximum of a vector	Accepts variable number of arguments
sample	Random sample	Also produces permutations
sum	Arithmetic sum	Also used for counting
var	Variance and covariance	Accepts vectors or matrices

Each of the statistical functions will return a missing value (**NA**) if any of the elements in the arguments to the functions contain a missing value. In S-PLUS, most of the functions accept an additional logical argument, called **na.rm**. If set to TRUE, the function in question will automatically remove all missing values before performing its calculations. As an alternative, subsetting operations, as described in Section 5.1.1, may be necessary. For example, suppose we wish to find the median of a vector that contains a missing value:

```
> tst <- c(5,9,8,NA,17,12,4,3,2,6)
> median(tst)
[1] NA
```

Under S-PLUS, all that is required is to pass na.rm=T to the median function:

```
> median(tst,na.rm=T)
[1] 6
```

Without the na.rm argument, a subsetting expression can be used to remove the missing values:

```
> median(tst[!is.na(tst)])
[1] 6
```

The quantile function calculates one or more quantiles for a numeric vector. (Quantiles are like percentiles, but instead of being in the range of 0 to 100, they are in the range of 0 to 1. Thus the 95th percentile corresponds to the 0.95 quantile.) When called with a single argument, quantile returns a vector of length 5 containing the minimum, the lower quartile (0.25 quantile), the median, the upper quartile (0.75 quantile), and the maximum. If a second argument is provided, each of its elements must be between 0 and 1, and quantile will return an object of the same length as the second argument containing the requested quantiles. For example, to get the default five-number summary for the length of automobiles in the auto.stats data set, we could use the statement

```
> quantile(auto.stats[,"Length"])
[1] 142.00 170.00 192.50 203.75 233.00
```

To get the minimum, maximum, and median for the same variable, the following statement could be used:

```
> quantile(auto.stats[,"Length"],c(0,1,0.5))
[1] 142.0 233.0 192.5
```

The cor function is typically called with a single matrix, the columns of which represent the values of two or more variables, and returns a matrix whose i, jth element is the correlation between the ith column (variable) and the jth column (variable) of the input matrix. Alternatively, cor can be called with two vectors or matrices. When called with two vectors, it returns the correlation between the vectors as a scalar. When either of the two arguments is a matrix, cor returns a matrix whose i, jth element is the correlation between the ith column in the first matrix and the jth column in the second matrix, where either i or j may always be equal to 1.

A closely related function is var, which calculates variances and variance-covariance matrices. Its behavior with various combinations of matrices and vectors is similar to that described for cor.

The function sample generates random samples or permutations of data. In its usual form, sample takes two arguments. The first represents either a vector of data to be sampled or a positive integer. In the latter case, the data to be sampled will consist of the integers from 1 through the positive integer used as an argument. The second argument to **sample**, **size**, is the size of the sample desired, which defaults to the length of the input vector if no value is specified. In addition, by default, sampling is done without replacement, so that once a value is chosen, it is not eligible to be chosen again. (This behavior can be overridden by passing **replace = T** to **sample**.) The combination of these default behaviors means that if you call **sample** with only one argument, the result will be a random permutation, either of the data or of the sequence of integers described above. For example, to create several different rearrangements of the numbers from 1 through 10, we could repeatedly call **sample** with the single argument **10**:

```
> sample(10)
[1]  1  3  7  9  8 10  6  4  2  5
> sample(10)
[1] 10  7  3  6  1  5  4  2  9  8
> sample(10)
[1]  5  7  2  1  8 10  6  3  4  9
```

The **sample** function can also be used to generate random numbers from discrete distributions; see Section 3.10.4.

In addition to calculating the sum of a vector of numbers, the **sum** function can also be used to count the number of elements in a vector that meet a certain condition, because logical values of **FALSE** and **TRUE** will be converted into numeric values of 0 and 1, respectively, when passed to the **sum** function. So to count the number of automobiles in the **auto.stats** data set whose gas mileage is greater than 20 miles per gallon, you could use the expression

```
> sum(auto.stats[,"Miles per gallon"] > 20)
[1] 36
```

The functions **min** and **max** return the minimum and maximum, respectively, of all the arguments passed to them. These two tasks are combined by the function **range**, which returns a numeric vector of length 2 containing the minimum (first position) and maximum (second position) of all the values passed to the function.

Two common statistics that are not built into S are skewness and kurtosis. These statistics, which are standardized moments around the mean, have historically been used to describe overall distribution characteristics, but are not as useful as more modern tools such as graphical displays. In addition, there is no clear consensus as to the exact definitions of these measures. The following definitions will calculate skewness and kurtosis values for a vector, using a widely used definition for these two quantities:

```
> skewness <- function(x)mean((x - mean(x))^3) / sqrt(var(x))^3
> kurtosis <- function(x)mean((x - mean(x))^4) / var(x)^2 - 3
```

3.10 Functions for Probability Distributions

S offers four types of functions relating to probability distributions, for each of several different distributions. In each case, the functions begin with one of the letters d, p, q, and r, which stand for *density, probability, quantile,* and *random sample,* respectively. The available distributions are shown in Table 3.9. For a particular distribution, the *name* shown in the table should be appended to one of the letters d, p, q, and r to get the name of the appropriate function. For example, to get density values for the Poisson distribution, use the function dpois; to get probability values from the Student's *t* distribution, use the function pt.

The first argument to any of the probability distribution functions beginning with the same letter plays the same role (described in the subsections below) regardless of the distribution desired. However, depending on the distribution, it may be necessary or desirable to pass additional parameters to the function. These additional parameters are listed, along with their defaults, in Table 3.9. If a dash (–) appears as a default value in the table, you must specify the parameter or an error will occur. If a dash appears under Range, there is no limitation on the value of the associated parameter.

Like most functions in S, the arguments to these functions may be vectors, in which case a vector will be returned, and the elements of the returned vector will correspond to the elements of the vector of arguments. The next four subsections describe the four basic groups of probability distribution functions.

3.10.1 Probability Density (d)

These functions evaluate the probability density function (p.d.f.) of the selected distribution. Their first argument is the ordinate (x value) at which the evaluation of the density is desired. Thus, values passed to these functions should have their first argument in the scale of observations derived from the distribution. If an argument outside of the range of support for the distribution is given (such as a negative value for the Gamma distribution), a value of 0 is returned.

3.10.2 Probabilities (p)

These functions return the probability that a sample from the specified distribution would be less than or equal to the first argument to the function; that is, they evaluate the cumulative distribution function (c.d.f.) of the distribution. Like the density, the argument should be in the scale of the observations. Note that these are not tail probabilities; to get the usual tail probabilities associated with statistical tests, the returned value should be subtracted from 1.

Table 3.9 Probability Distributions Available in S

Distribution	Name	Parameters	Defaults	Range
Beta	*beta*	shape1	-	shape1 > 0
		shape2	-	shape2 > 0
Binomial	*binom*	size	-	size > 0 (integer)
		prob	-	0 < prob < 1
Cauchy	*cauchy*	location	0	-
		scale	1	scale > 0
Chi-squared	*chisq*	df	-	df > 0
Exponential	*exp*	rate	1	rate > 0
F	*f*	df1	-	df1 > 0
		df2	-	df2 > 0
Gamma	*gamma*	shape	-	shape > 0
Geometric	*geom*	prob	-	0 < prob < 1
Lognormal	*lnorm*	meanlog	0	-
		sdlog	1	sdlog > 0
Logistic	*logis*	location	0	-
		scale	1	scale > 0
Negative binomial	*nbinom*	size	-	size > 0 (integer)
		prob	-	0 < prob < 1
Normal	*norm*	mean	0	-
		sd	1	sd > 0
Poisson	*pois*	lambda	-	lambda > 0
Student's *t*	*t*	df	-	df > 0
Uniform	*unif*	min	0	min < max
		max	1	max > min
Weibull	*weibull*	shape	-	shape > 0

3.10.3 Quantiles (q)

The quantile functions accept a probability as their first argument, and return a value such that the probability of obtaining a value less than or equal to the returned value is the probability given as the argument. Thus, the first argument to these functions must be in the range of 0 to 1. If a value outside this range is used, the function will return a missing value (**NA**), and an error message may be printed. The quantile functions are the inverse of the probability functions described in the preceding subsection.

3.10.4 Random Samples (r)

The random sample functions generate a vector of random samples from the specified distribution. (Actually, like all computer-generated random numbers, these are really pseudo-random numbers.) An internal uniform random number generator is the basis of all these functions; each time you call one of these functions, a vector of values, stored in an object named .Random.seed in your .Data directory, is updated. Thus, each time you call any of the random number generators, you will get a different random sample, up to the limit of the period of the generators, which should be sufficiently large to serve most purposes. If you wish to repeat a specific sequence of random numbers, you will need to reset .Random.seed to the value it had when you first called it. For example, here are three calls to the random number generator rgamma, for five random numbers, each with a shape parameter equal to 2:

```
> rgamma(5,2)
[1] 1.0899313 1.1206256 5.7983772 0.3070747 1.9806954
> rgamma(5,2)
[1] 0.2919759 1.8956504 3.5541317 0.2697249 2.1311822
> rgamma(5,2)
[1] 0.3723883 3.9084747 3.5885586 0.8574431 1.1435639
```

Notice that a different sequence is obtained each time. If the value of .Random.seed is stored, then reset, we can reproduce any desired sequence:

```
> save.seed <- .Random.seed
> rgamma(5,2)
[1] 1.010075 1.783897 2.917057 3.515872 1.942692
> rgamma(5,2)
[1] 1.4761555 2.7322962 1.6731025 1.3708507 0.7550122
> .Random.seed <- save.seed
> rgamma(5,2)
[1] 1.010975 1.783897 2.917057 3.515872 1.942692
```

Note that after the value of .Random.seed is reset, the first sequence of random numbers is reproduced. You can also reset the random number generator to a repeatable value using the function set.seed.

One convenient feature of the random number generators is their treatment of vector-valued arguments. Specifically, the elements of a vector-valued argument will be used in turn, recycling if necessary. For example, to generate 100 samples from a normal distribution with mean 0 for most observations, but with mean 1 for every fifth observation, we could use the following statement:

```
> rnorm(100,mean=c(0,0,0,0,1))
```

Because there are only five values in the mean= argument, they will be reused in order to produce the sample of 100. No error or warning message is printed if there are too few or too many values in a vector-valued argument to these functions; they are simply recycled silently.

In addition to the functions described above, the function sample can be used to generate random values from discrete distributions. The first argument to sample is the vector of possible discrete values, or a scalar integer, in which case a vector of integers from 1 to the argument is used. The prob argument specifies a vector of probabilities whose length is equal to the number of different discrete values, and the size argument specifies the sample size desired. (If the probabilities specified do not add up to 1, they will be silently normalized.) When you use **sample** to generate random numbers in this way, the argument **replace=T** should always be used. For example, to generate 50 integers from the set of 1, 2, 3, 4, and 5, with probabilities 0.15, 0.2, 0.3, 0.2, and 0.15, respectively, we could use the following statement:

```
> sample(5,prob=c(.15,.2,.3,.2,.15),size=50,replace=T)
 [1] 3 2 3 5 3 2 4 3 3 5 4 4 2 5 3 1 2 3 5 4 5 3 2 5 3 2 1 3 1 3
[31] 3 3 2 3 3 5 5 4 1 2 3 1 4 3 3 5 1 2 3 2
```

3.11 Functions for Categorical Variables

A categorical variable is one that takes on only a small number of values, each representing a different level of some measurement. In S, categorical variables are important because, among other uses, they are used to represent factors in statistical models, discussed in Chapter 10. Table 3.10 lists some of the functions available for categorical variables.

Table 3.10 Functions for Categorical Variables

Name	Function
codes	Converts factor levels to integers
cut	Creates categories out of a continuous variable
factor	Creates a factor from a categorical variable
levels	Displays or sets the levels of a factor or categorical variable
pretty	Creates convenient break points for a categorical variable
split	Breaks up an array according to the value of a categorical variable
table	Counts the number of observations cross-classified by categories

The function **cut** creates a categorical variable from a continuous variable, by assigning the values of the variable into nonoverlapping ranges and creating a new variable based on the range in which each value lies. Either you can specify the number of groups you want, in which case **cut** will create a set of equal-sized ranges, or you can explicitly provide the values which define the ranges. There are two required arguments to **cut**. The first is the vector of values to convert to a categorical variable. The second is either a scalar indicating the number of ranges to create, or a vector containing the values that define the boundaries of the ranges (**breaks**). Any value outside these ranges will result in a missing value (**NA**). Thus, if you provide a vector of values, you should be sure that the first value is smaller than any of the data values, and that the last value is larger than any of the data values. There will be one less category in the vector created by **cut** than there are values in the vector of break points provided. For example, suppose we wish to divide the automobile data set into three groups, based on the weight of the automobiles. If we define a small car as one that weighs 2500 pounds or less, and a large car as one that weighs more than 3500 pounds, we can create a variable called `cargrp` as follows:

```
> cargrp <- cut(auto.stats[,"Weight"],c(0,2500,3500,5000))
> cargrp
 [1] 2 2 2 2 1 2 2 3 3 1 2 3 2 3 3 3 3 1 3 2 2 2 2 1 1 1 2 1 3 3 3
[31] 1 1 2 1 1 3 3 3 1 2 3 3 3 2 2 3 2 2 3 2 2 3 2 1 1 1 2 2 3 2
[61] 2 2 2 2 1 1 1 1 2 1 1 1 1 2
attr(, "levels"):
[1] "   0+ thru 2500" "2500+ thru 3500" "3500+ thru 5000"
```

The value of 5000 was chosen because it is larger than the maximum value of `Weight` for any car in the data set. Notice that the `cut` function automatically provided values for the levels attribute of `cargrp`.

When you specify an integer as the second argument to `cut`, it will divide the range of values encountered into equally spaced intervals. If the data is not evenly spread over the entire range of values, some of the levels defined by `cut` may not contain many observations. For example, suppose we wish to divide the states in the `state.x77` data set into three groups, based on their areas. As a first try, we might use the following statements:

```
> size.grp <- cut(state.x77[,"Area"],3)
> size.grp
 [1] 1 3 1 1 1 1 1 1 1 1 1 1 1 1 1 1 1 1 1 1 1 1 1 1 1 1 1 1 1 1
[31] 1 1 1 1 1 1 1 1 1 1 1 1 2 1 1 1 1 1 1 1
attr(, "levels"):
[1] "Range 1" "Range 2" "Range 3"
```

Notice that there is only one observation in each of ranges 2 and 3. To create groups with equal numbers of observations (as opposed to equal interval widths), you can use the `quantile` function to choose the appropriate break points:

```
> size.grp <- cut(state.x77[,"Area"],
+                  c(0,quantile(state.x77[,"Area"],(1:3)/3)))
> size.grp
 [1] 2 3 3 2 3 3 1 1 2 2 1 3 2 1 2 3 1 2 1 1 1 2 3 2 2 3 3 3 1 1
[31] 3 2 2 3 1 2 3 2 1 1 3 1 3 3 1 1 2 1 2 3
attr(, "levels"):
[1] "   0.00+ thru  42528.67" "42528.67+ thru  69180.33"
[3] "69180.33+ thru 566432.00"
```

The value of 0 was explicitly added to the break points to include the minimum value of `Area` in the first range.

You can specify a character vector of levels to be stored with a variable by passing the optional `labels=` argument to `cut`:

```
> cargrp <- cut(auto.stats[,"Weight"],c(0,2500,3500,5000),
+               labels=c("Small","Medium","Large"))
```

```
> cargrp
[1] 2 2 2 2 1 2 2 3 3 1 2 3 2 3 3 3 1 3 2 2 2 2 1 1 1 2 1 3 3 3
[31] 1 1 2 1 1 3 3 3 1 2 3 3 3 2 2 3 2 2 3 2 2 3 2 1 1 1 2 2 3 2
[61] 2 2 2 2 1 1 1 1 2 1 1 1 1 2
attr(, "levels"):
[1] "Small"  "Medium" "Large"
```

The same effect could be achieved by using the **levels** function to set the levels attribute of a categorical variable:

```
> levels(cargrp) <- c("Small","Medium","Large")
```

By default, the printing routines do not automatically use the levels attribute when an object is printed, as in the previous example, where the values of `cargrp` were displayed as 1, 2, and 3. However, if you convert a variable to a factor, then the levels will be used when the object is displayed:

```
> cargrp <- factor(cargrp)
> cargrp
[1] Medium Medium Medium Medium Small  Medium Medium Large
[9] Large  Small  Medium Large  Medium Large  Large  Large
[17] Small  Large  Medium Medium Medium Medium Small  Small
[25] Small  Medium Small  Large  Large  Large  Small  Small
                        . . .
```

The term *factor* comes from the use of factors in statistical modeling, discussed in Chapter 10. Keep in mind that once a variable is made a factor, references to its values must be made using the values of the levels attribute. For example, with `cargrp` converted to a factor, we could extract information for small cars only using the expression `auto.stats[cargrp == "Small",]`, not `auto.stats[cargp == 1,]`, as would be the case if `cargrp` were not a factor. To convert a factor's value back to integer levels, use the function **codes**. When you create a factor, missing values by default continue to be missing, and thus are not treated as a valid level of the factor. If you wish to include missing values as a valid level of a factor, the function **na.include** can be used.

Because factors have a fixed number of levels, they are sometimes internally converted to codes before being used in other operations. For example, if we try to combine the `cargrp` factor with a fourth classification, `Tiny`, using the **c** function, we get the following result:

```
> newgrp <- c(cargrp,"Tiny")
> newgrp
                        . . .
[57] "2"    "2"    "3"    "2"    "2"    "2"    "2"    "2"
[65] "1"    "1"    "1"    "1"    "2"    "1"    "1"    "1"
[73] "1"    "2"    "Tiny"
```

In cases such as these you can coerce the factor into a vector of character strings before the concatenation, using the **as.character** function.

The function **table** produces a contingency table, which is a vector or array containing the number of observations that have particular values of one or more variables. When just one argument is passed to `table` it counts the number of values for each level of a variable, producing a vector of length equal to the number of distinct values in its argument and using the levels attribute of its argument to label the different categories:

```
> table(cargrp)
  Small Medium Large
     23     31    20
```

With more than one argument to `table`, observations are divided into cells— that is, distinct combinations of each of the variables passed to `table`—and are then counted. A multi-way array is formed, with counts representing the number of observations in each cell defined by the arguments. For example, we could create a contingency table for the `cargrp` variable and the 1978 repair records as follows:

```
> table(cargrp,auto.stats[,"Repair (1978)"])
         1 2  3 4 5
  Small  0 0  6 8 8
  Medium 2 4 15 3 3
  Large  0 1  0 7 0
```

The left-hand side of the table lists the levels of the `cargrp` variable; the top of the table lists the values of the repair variable.

The function `split` breaks up one object based on the value of a categorical variable, producing a list with the pieces of the object as its elements. It is especially useful when applied to a data frame, because in that case each element of the list will be a data frame containing all the variables in the original object. The `split` function is also useful when producing plots that take as input a list of values (such as `boxplot`), but, in general, the programming techniques described in Chapter 7 are usually more efficient for processing groups of data.

3.12   Functions for Character Manipulation

S provides a small number of functions to deal with character values, and relies on the operating system interface to other character-handling routines. (Functions to access the operating system are discussed in detail in Section 8.8.) The S functions for character manipulation are summarized in Table 3.11.

One of the most basic character-handling commands is `paste`. The `paste` function allows you to combine character strings in a variety of ways. Like most S functions, it is smart about handling vector and scalar arguments. As a simple example, suppose we wish to generate variable names that consist of the string "var" followed by the numbers from 1 to 10. We can call `paste` with a single character string as its first argument and a vector of values as its second argument:

Table 3.11 Functions for Character Manipulation

Name	Function
abbreviate	Generate abbreviations of character values
cat	Display values on screen or send to file
grep	Search for patterns in characters
nchar	Number of characters in string
paste	Combine values into character strings
substring	Extract parts of character values

```
> paste("var",1:10)
 [1] "var 1"  "var 2"  "var 3"  "var 4"  "var 5"  "var 6"
 [7] "var 7"  "var 8"  "var 9"  "var 10"
```

Note that by default **paste** inserts spaces between all its arguments before pasting them together. It also converts numeric values to character values as needed before combining them.. To control what **paste** inserts between values, you can use the optional argument **sep=**. For example, you can suppress the space with the first **paste** command shown below, or specify a period with the second invocation:

```
>  paste("var",1:10,sep="")
 [1] "var1"  "var2"  "var3"  "var4"  "var5"  "var6"  "var7"
 [8] "var8"  "var9"  "var10"
> paste("var",1:10,sep=".")
 [1] "var.1"  "var.2"  "var.3"  "var.4"  "var.5"  "var.6"
 [7] "var.7"  "var.8"  "var.9"  "var.10"
```

You can also use **paste** to combine the strings in a vector into a single character value. The optional argument **collapse=** allows you to specify a character string to be inserted between the component parts; as with **sep**, a null string ("") will insert nothing between the pieces. For example, we could collapse the contents of the object **letters** as follows:

```
> letters
 [1] "a" "b" "c" "d" "e" "f" "g" "h" "i" "j" "k" "l" "m" "n" "o"
[16] "p" "q" "r" "s" "t" "u" "v" "w" "x" "y" "z"
> paste(letters,collapse="")
[1] "abcdefghijklmnopqrstuvwxyz"
> paste(letters,collapse="_")
[1] "a_b_c_d_e_f_g_h_i_j_k_l_m_n_o_p_q_r_s_t_u_v_w_x_y_z"
```

The use of the **paste** function in conjunction with the **cat** function to produced aligned displays is discussed in Section 8.5.

The function **nchar** returns the number of characters in a character string, which is often called the length of the character string. (Note that the meaning of *length* in this case differs from the usual meaning of the length of an S object, which is the number of elements in the object.) If the argument to **nchar** is a vector of character strings, it will return a vector of lengths.

The function **substring** returns parts of a character string through the use of the arguments **first** and **last**, representing the ordinal positions of the first and last characters of the extracted substring. If the **last** argument is eliminated, the substring from the **first** character to the end of the string is returned. By using a vector for either or both of these arguments, a vector containing the pieces of a character string can easily be created. For example, to undo the collapsing of the **letters** vector, we could use the following call to **substring**:

```
> clett <- paste(letters,collapse="")
> clett
 [1] "abcdefghijklmnopqrstuvwxyz"
> nc <- nchar(clett)
> substring(clett,1:nc,1:nc)
 [1] "a" "b" "c" "d" "e" "f" "g" "h" "i" "j" "k" "l" "m" "n" "o"
[16] "p" "q" "r" "s" "t" "u" "v" "w" "x" "y" "z"
```

The vector arguments are matched up in order so that the first element of the output is substring(clett, 1, 1), the next is substring(clett, 2, 2), and so on.

The function **abbreviate** will attempt to form shortened versions of character strings by eliminating whitespace, lowercase vowels, lowercase consonants, punctuation, and uppercase letters, in that order. This can be useful if a printed or graphical display does not have sufficient space to display character strings in their full form. The first argument to **abbreviate** is a vector of character values to be abbreviated; the second argument is a suggested minimum length to be attained, set by default to 4. It cannot be guaranteed that all abbreviations will be as small as the suggested minimum length, because the output from **abbreviate** will always create a unique abbreviation for each unique value in the input vector. As an example of the use of **abbreviate**, we can create a list of shortened names for the countries in the **saving.x** data set:

```
> abbreviate(dimnames(saving.x)[[1]])
Australia Austria Belgium Bolivia Brazil Canada  Chile Taiwan
  "Astrl"   "Austr" "Blgm"  "Bolv"  "Brzl" "Cand" "Chil" "Tawn"

Colombia Costa Rica Denmark Ecuador Finland France Germany
  "Clmb"    "CstR"    "Dnmr"  "Ecdr"  "Fnln"  "Frnc" "Grmn"

Greece Guatemala Honduras Iceland  India Ireland  Italy
 "Grec" "Gtml"    "Hndr"   "Icln"  "Indi" "Irln"  "Itly"

 Japan  Korea Luxembourg  Malta Norway Netherlands
"Japn" "Kore" "Lxmb"      "Malt" "Nrwy" "Nthr"
             . . .
```

The S function **grep**, like its UNIX counterpart, searches for a regular expression.[2] The first argument to **grep** is the pattern of interest; the second argument is the vector of character values that should be searched for the pattern. The function returns the index (or indices) of those elements within the vector that contain the pattern. For example, when using the function **objects**, introduced in Section 2.4, the argument **where** can be used to specify the index of a directory in the search list for which its contents should be listed. Suppose we wish to list the functions in the statistics library. We could display the search list and determine the number in that fashion, or we could use the function **grep** as follows:

```
> objects(where=grep("statistics",search()))
```

This technique works because the function **search** returns a vector of character values representing the search directories. To find the positions of a character within a single character string, the function **grep** is not directly usable, because it expects its second argument to be a vector of character values. Thus, the following statement will not give the expected result:

```
> grep("a","cat")
[1] 1
```

The desired result can be achieved by using the **substring** function to create a vector containing the individual characters, then passing it to **grep** as follows:

```
> grep("a",substring("cat",1:3,1:3))
[1] 2
```

A series of such manipulations can be combined to break apart character strings in more complex ways. For example, we could find the "tail" of a filename— that is, the portion of the filename after its last slash (/)—using the following statements:

```
> filename <- "/usr/local/lib/test.data"
> nc <- nchar(filename)
> sl <- grep("/",substring(filename,1:nc,1:nc))
> lsl<- length(sl)
> substring(filename,first = sl[lsl] + 1)
[1] "test.data"
```

Since **grep** returns a vector of indices, the expression **sl[lsl]** will be the index of the last slash encountered.

---

[2] The regular expressions used by the **grep** command are similar to those used by the UNIX command **egrep**. See your local UNIX documentation for more information. Under MS-DOS and Windows, wildcards are limited to ? for a single character and * for any sequence of characters.

Exercises

1. The empirical rule states that approximately 68% of data from a normal distribution with a mean of 0 and a standard deviation of 1 will have an absolute value less than 1. Use the **sum** and **rnorm** functions to find the proportion of 1000 random normal variables whose absolute value is less than 1. Repeat several times and see how widely the results vary.

2. When a matrix is singular—that is, when an inverse cannot be computed— a matrix known as the generalized inverse is often used instead of the true inverse. For a square matrix X, the generalized inverse, X^+, can be calculated from the quantities U, V, and D obtained from the singular value decomposition of X (Section 3.6.5) as

$$X^+ = VD^+U'$$

where D^+ is a diagonal matrix composed of the reciprocal of the corresponding nonzero elements of D, and zeroes elsewhere. Write an expression to calculate the generalized inverse of a singular square matrix. (You can generate a singular matrix by forming the quantity $X'X$ (x%*%t(x)) for a matrix X that has more rows than columns.) You may find the **ifelse** function (Section 7.4) useful in calculating D^+. Make sure to treat very small elements of D as if they were zero.

4 Introduction to Graphics in S

4.1 Overview

The S language provides a wide collection of functions for producing graphics, both printed and displayed on computer screens. There are two basic types of graphics functions in S: high-level plotting commands, which will produce a complete plot with a single function invocation, and low-level plotting commands, which will add additional features to an existing plot. In addition, the function **par** allows you to set or query a wide range of graphics parameters, which control the way in which the actual graphics functions operate. In this chapter, the high-level plotting commands are presented along with examples of their uses. These commands allow you to produce a complete plot with just a single function call. To customize your graphs, you can use the low-level function calls and graphics parameters described in Chapter 9. Most graphics tasks in S start with a call to a high-level function, so the material in this chapter should help you get started producing graphical output very quickly.

Before any graphics commands can be issued, an appropriate graphic device must be selected, and the corresponding graphics function called. For interactive use, you'll probably want to display your graphs on the screen of your computer. On UNIX machines, under X Windows, you can use either the **motif()** or the **openlook()** driver in S-PLUS, or **x11()** in S. For S-PLUS running under Windows, use the **win.graph()** driver, and for S-PLUS running under DOS without Windows, use either **vga()**, **ega()**, or **hercules()**, depending on the type of graphics card that is installed in your computer. In addition, if you are using S-PLUS, Section 9.9 explains how you can manipulate more than one graphics device at a time, for example to view several different graphs simultaneously, each occupying an entire graphics window.

To load a graphics driver, simply call the appropriate function with no arguments. For example, to create a graphics window for S-PLUS under Windows, type:

```
> win.graph()
```

The help file for the `win.graph` function will have information on available arguments to the `win.graph` function, although you will rarely need to use any arguments when you invoke a device for interactive use.

When you are running S-PLUS under UNIX, both the `motif` and `openlook` functions provide a window with a button, which can be used to produce printed output easily; under other environments you can send the output from your graphics commands to a printer by invoking the appropriate device function for the type of printer you're using. Under Windows, the command `win.printer()` can be used to direct the output of subsequent commands to your printer; under other operating systems, the possibilities include `postscript()` and `hpgl()`. When you use these devices, you can have the output sent either to a file or directly to the printer by passing appropriate arguments to the function in question. For more details, see the online documentation for these functions.

You can determine what devices are available in your version of S by typing

```
> help(Devices)
```

Since new devices are often added to new versions of S, this is the most effective way to find out exactly what is available on your system. In addition, if you have specialized graphics devices at your location, additional locally written device functions may be available. Check with your local S administrator to see if this is possible.

Device functions that do not directly produce an image on the computer screen generally write their output to a location in computer memory called a buffer before actually writing them to a file or sending them as input to some external command, such as a printing command. When you terminate your S session, any information in the buffer is sent to the appropriate file or command. However, if you wish to use the output before your S session is over (for example, printing a file produced by your graphics commands), you should use the function `graphics.off()` to flush any information in the buffer and deactivate the graphics device. If you then decide to produce more plots in the same session, you'll need to reinvoke the graphics function once again. Otherwise, a single call to a graphics function is all you need within a single S session.

4.2 High-Level Plotting Commands

Table 4.1 lists the names and brief descriptions of some of the high-level graphics commands found in S. As mentioned previously, once a device function has been appropriately invoked, a single call to one of these routines will produce a complete plot.

Table 4.1 High Level Plotting Functions

Function	Description	Example Figure
barplot	Vertical or horizontal bar graph	4.1(a)
boxplot	Side-by-side boxplots	4.1(b)
contour	Contour plot	4.1(c)
coplot	Separate plots for different ranges	4.1(d)
dotchart	Displays values based on positions of dots	4.2(a)
faces	Chernoff faces for multivariate data	4.2(b)
hist	Histogram	4.2(c)
pairs	All possible pairs of scatterplots	4.2(d)
persp	3-D perspective plots of grids	4.3(a)
pie	Pie charts	4.3(b)
plclust	Plots of cluster trees from hclust	4.3(c)
plot	Scatterplot or line plot	4.3(d)
qqnorm	Normal probability plot	4.4(a)
qqplot	Quantile-quantile plots	4.4(b)
tsplot	Plotting of time series	4.4(c)
usa	Map of United States	4.4(d)

Figures 4.1 through 4.4 provide a simple illustration of each of the high-level plotting routines shown in Table 4.1. The statements that produced these plots are presented in Section 4.5.

4.3   Arrangement of Plots on the Page or Screen

A complete discussion of graphics parameters that allow you to control the details of the way your graphic output will look will be presented in Section 9.2. However, two graphics parameters that are very useful in arranging multiple plots are worth mentioning here. They are `mfrow` and `mfcol`, and each of them defines a rectangular array of plots within a page (or screen) of graphics output. Values for both of these parameters consist of a vector of length 2, giving the number of rows and columns in the array of plots desired; `mfrow` then causes the array to be filled with your graphics output row by row, whereas `mfcol` fills the array in a column-by-column fashion. When a value of one of these parameters is in effect, subsequent calls to high-level plotting routines will cause graphics output to the next position in the array defined by either the `mfrow` or the `mfcol` parameter. To set either of these parameters, you can use the function `par`, as in the following example, which produces histograms and scatterplots versus vehicle weight for several variables in the `auto.stats` data set. The plots, shown in Figure 4.5, are arranged in a 3-by-2 array, and output one row at a time; the title of the rightmost plot in each row shows the order in which the rows were output.

Figure 4.1   Graphics Examples, Part 1

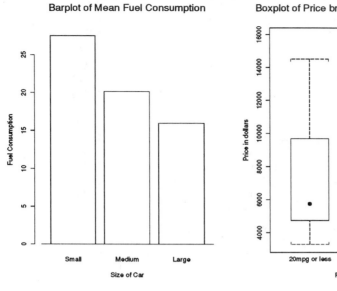

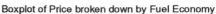

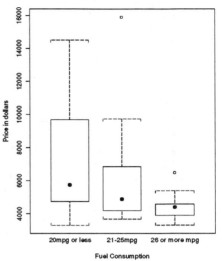

(a) Barplot

(b) Boxplot

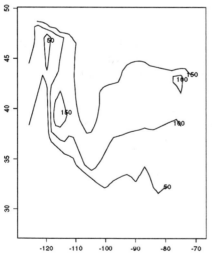

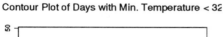

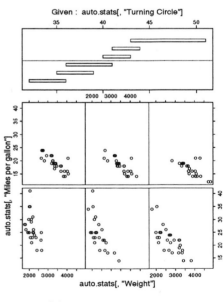

(c) Contour Plot

(d) Conditioning Plot

Figure 4.2   Graphics Examples, Part 2

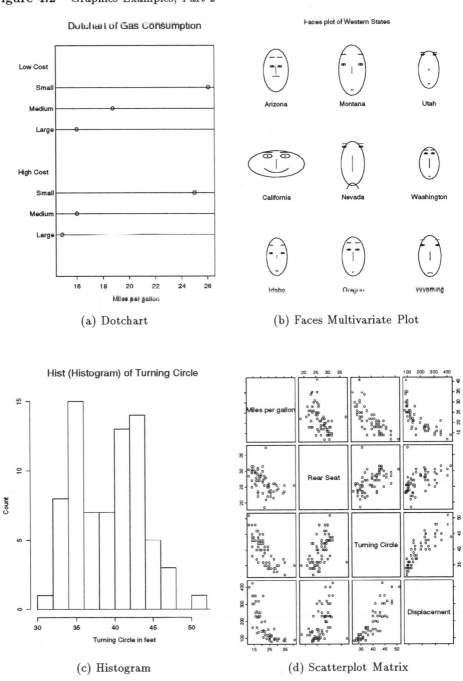

(a) Dotchart

(b) Faces Multivariate Plot

(c) Histogram

(d) Scatterplot Matrix

Figure 4.3   Graphics Examples, Part 3

Persp plot of sin(x) * cos(y)

Pie Chart of Sizes of cars in auto.stats

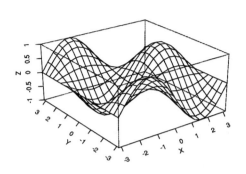

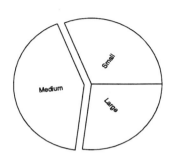

(a) 3-D Perspective Plot

(b) Pie Chart

Piclust of Western States

Plot of Sin(theta) vs. theta

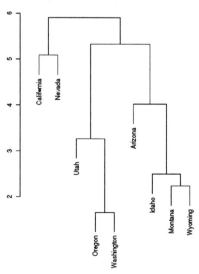

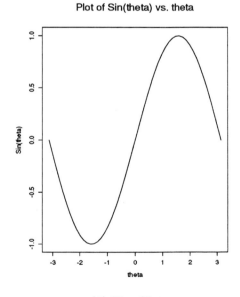

(c) Cluster Tree Plot

(d) Line Plot

Figure 4.4   Graphics Examples, Part 4

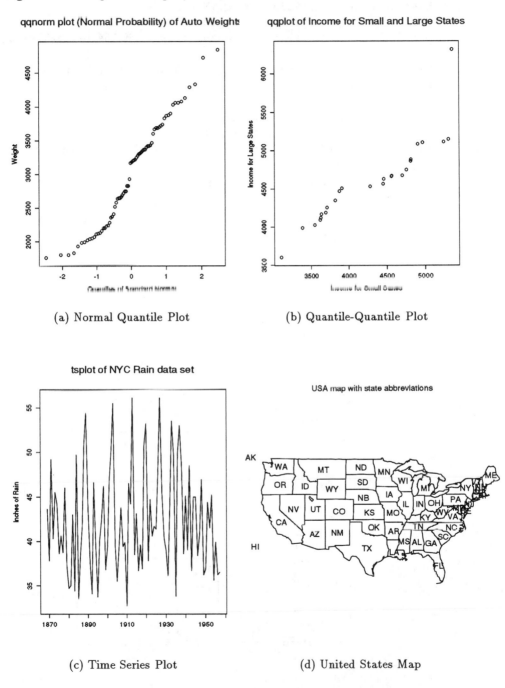

(a) Normal Quantile Plot

(b) Quantile-Quantile Plot

(c) Time Series Plot

(d) United States Map

```
> par(mfrow=c(3,2))
> hist(auto.stats[, "Miles per gallon"],
+ main = "Histogram of Gas Mileage")
> plot(auto.stats[, "Weight"], auto.stats[, "Miles per gallon"],
+ main = "First")
> hist(auto.stats[, "Trunk"], main = "Histogram of Trunk")
> plot(auto.stats[, "Weight"], auto.stats[, "Trunk"],
+ main = "Second")
> hist(auto.stats[, "Displacement"],
+ main = "Histogram of Displacement")
> plot(auto.stats[, "Weight"], auto.stats[, "Displacement"],
+ main = "Third")
```

To restore graphics output to a single plot per page, use the command

```
> par(mfrow=c(1,1))
```

For more complex layouts, such as different numbers of plots in each row or column, see Sections 9.3 and 9.10.

Figure 4.5   Multiple Plots on a Page

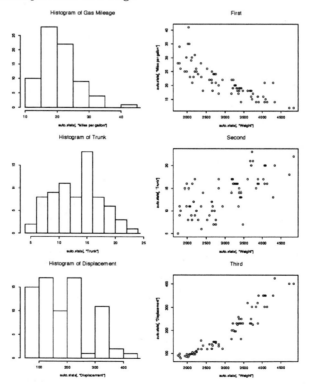

4.4 Interacting with Plots

If the graphics device you use is equipped for one, you can use a mouse or other pointing device to interact with your plots, either identifying and labeling points on a plot or finding the user coordinates of any location in the plotting region. The two functions that implement these methods are **identify** and **locator**.

The first two arguments to the function **identify** are the same values for **x** and **y** that produced the plot with which you are interacting. You can then use the pointing device to select the point of interest. By default, **identify** will then write the subscript of the value identified on the graph near the position of the point, on the side of the point at which the pointing device was activated. By passing an optional third argument to **identify**, you can specify labels of your choice to be placed on the plot; the first component of the **dimnames** attribute is often a good choice. For example, using the **auto.stats** data set, a plot of length versus rear seat clearance shows one point fairly distant from the others. To print an appropriate label next to that point, we could use the following statements in conjunction with the pointing device:

```
> plot(auto.stats[,"Length"],auto.stats[,"Rear Seat"],
+       xlab="Length",ylab="Rear Seat",
+       main="Plot of Rear Seat Clearance vs. Length")
> identify(auto.stats[,"Length"],auto.stats[,"Rear Seat"],
+       labels=dimnames(auto.stats)[[1]])
```

Notice that the first two arguments to **identify** are identical to the corresponding arguments to **plot**. The plot, with the outlying point identified as "Volks Dasher," is shown in Figure 4.6.

By default, the label is printed 0.5 character width from the point selected; this value can be changed with the **offset** argument. Alternatively, labels can be printed at the point in question by setting the argument **atpen** to **FALSE**. The printing of the labels on the graph can be suppressed by setting the argument **plot** to **FALSE**. In addition to the printing of the label, **identify** returns a vector containing the subscripts of all the points that were identified. This returned vector is useful for extracting or eliminating subsets of points from matrices or vectors using the subsetting techniques described in Section 5.1.1. If no point on the plot is within 0.5 inch of the position selected by the pointing device, a warning message will be printed. The distance at which the message is printed can be controlled using the **tolerance** argument to **identify**.

In addition to the arguments mentioned above, **identify** also accepts graphics parameters. Some parameters that might be useful when using **identify** include **srt** for rotating the printed labels, **cex** for increasing the size of the characters used in the labels, and **adj** for controlling the justification of the labels.

Although **identify** is useful for labeling points, there are times when interactions with graphics output are not related to points already on the plot. For example, you may be trying to find a good location for a legend or some

Figure 4.6   Plot with Outlier Identified with `identify`

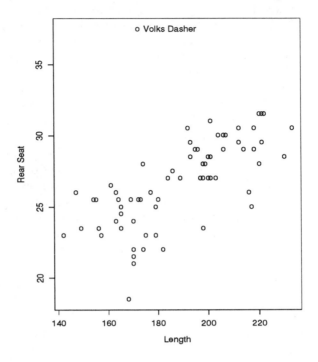

The type="n" argument to the plot function tells it to set up axes, without
actually doing any plotting.

explanatory text, or you may want to interactively draw a plot by specifying the
locations of the points directly on the graph. The `locator` function returns a
list with components named **x** and **y**, containing x and y values corresponding to
the user coordinates of the points selected. By changing the value of the argument **type** from its default of **"n"** (none), you can plot points (**"p"**), lines (**"l"**),
or both points and lines (**"b"**) at the locations selected by the pointing device.
The optional argument **n** informs `locator` of the maximum number of points to
locate; after that number of points, `locator` will return the list described above
with no further input from the pointing device. Thus, for example, you could
plot a line through 10 points by "connecting the dots" of the chosen points using
the following statements:

```
> plot(1:10,1:10,type="n")
> lines(locator(10))
```

The **type="n"** argument to the **plot** function tells it to set up axes, without
actually doing any plotting.

    For the specific details of using the pointing device associated with your
graphics device, consult the online help file for the device in question.

4.5   Commands for Graphics Examples

barplot

```
cargrp <- factor(cut(auto.stats[,"Weight"],c(0,2500,3500,5000),
            labels=c("Small","Medium","Large")))
barplot(tapply(auto.stats[, "Miles per gallon"], cargrp, mean),
        names = levels(cargrp), ylab = "Fuel Consumption",
        xlab = "Size of Car",
        main = "Barplot of Mean Fuel Consumption")
```

---

boxplot

```
egrp <- cut(auto.stats[,"Miles per gallon"],c(0,20,25,50),
            labels = c("20mpg or less", "21-25mpg","26 or more mpg"))
boxplot(split(auto.stats[, "Price"], egrp), xlab =
        "Fuel Consumption", ylab = "Price in dollars", main =
        "Boxplot of Price broken down by Fuel Economy")
```

---

contour

```
frost <- interp(state.center$x,state.center$y,state.x77[,"Frost"])
contour(frost)
title(main="Contour Plot of Days with Min. Temperature < 32")
```

---

coplot

```
coplot(auto.stats[,"Miles per gallon"]~auto.stats[,"Weight"] |
        auto.stats[,"Turning Circle"])
```

---

dotchart

```
sgroup <- cut(auto.stats[, "Weight"], 3,
            c("Small", "Medium", "Large"))
pgroup <- cut(auto.stats[, "Price"], 2,
            c("Low Cost", "High Cost"))
tbl <- tapply(auto.stats[,"Miles per gallon"], list(pgroup, sgroup),
            mean)
dotchart(tbl,levels(category(col(tbl),
        labels = levels(sgroup)))[col(tbl)],
        category(row(tbl),labels = levels(pgroup)),
        xlab="Miles per gallon",
        main = "Dotchart of Gas Consumption")
```

---

faces

```
west <- c(3, 5, 12, 26, 28, 37, 44, 47, 50)
west.states <- state.x77[west,  ]
faces(west.states,c(8, 1, 2, 3:7),dimnames(west.states)[[1]],
        head="Faces plot of Western States")
```

hist

```
hist(auto.stats[,"Turning Circle"],ylab="Count",
        xlab="Turning Circle in feet",
        main="Hist (Histogram) of Turning Circle")
```

pairs

```
pairs(auto.stats[,c(2,6,10,11)])
```

persp

```
pts <- seq(-pi,pi,len=20)
z <- outer(pts,pts,function(x,y)sin(x)*cos(y));
persp(x=pts,y=pts,z)
title(main="Persp plot of sin(x) * cos(y)")
```

pie

```
cargrp <- factor(cut(auto.stats[,"Weight"],c(0,2500,3500,5000),
                labels=c("Small","Medium","Large")))
pie(table(cargrp), names = levels(cargrp), explode = 2,
        main="Pie Chart of Sizes of cars in auto.stats")
```

plclust

```
west <- c(3,5,12,26,28,37,44,47,50)
west.states <- state.x77[west,  ]
west.states <- sweep(west.states, 2, apply(west.states, 2,
function(x) sqrt(var(x))), "/")
west.cl <- hclust(dist(west.states))
plclust(west.cl, labels = dimnames(west.states)[1],
        main="Plclust of Western States")
```

plot

```
pts <- seq(-pi,pi,len=50)
plot(pts,sin(pts),type="l",xlab="theta",
        ylab="Sin(theta)",main ="Plot of Sin(theta) vs. theta")
```

qqnorm

```
qqnorm(auto.stats[, "Weight"], ylab = "Weight",
       main="qqnorm plot (Normal Probability) of Auto Weight")
```

qqplot

```
sz <- cut(state.x77[,"Area"],
          c(0,median(state.x77[,"Area"]),max(state.x77[,"Area"])))
inc <- split(state.x77[,"Income"],sz)
qqplot(inc[[1]], inc[[2]], xlab = "Income for Small States",
       ylab="Income for Large States", main =
       "qqplot of Income for Small and Large States")
```

tsplot

```
tsplot(rain.nyc1,main="tsplot of NYC Rain data set",
       ylab="Inches of Rain")
```

usa

```
usa()
text(state.center$x,state.center$y,state.abb)
title(main="USA map with state abbreviations")
```

Exercises

1. Use the command **help(Devices)** to determine what devices are supported by your version of S, then find out (from your manual, your system administrator, or other users) which ones are available on your computer system.

2. If your computer is equipped to do so, use the **locator** function to interactively produce a legend or title on a plot. Note that the **locator** function returns the x and y coordinates of the point you choose. How could you use these when you were producing a plot in a noninteractive setting?

3. Follow the example of a perspective plot (**persp**), given above, to plot the function $(2 + sin\ x)(cos\ 2y)$ with x and y both ranging from -3 to 3. Experiment with using different values of the **eye** argument to **persp**, which determines the angle at which the plot is viewed.

4. Experiment with the **which** argument of the **faces** command to change the features on the faces to correspond to different variables, using either the Western states example in the preceding section or a data set of your choice. Which feature(s) seem to be most effective in communicating differences to you?

5 Subsetting and Reshaping Data

When you are dealing with a data set or a matrix, there are times when it is convenient or necessary to use only a part of the data for a calculation or analysis. Sometimes this process is defined by choosing some columns by number or name, for example if it is desired to repeat a calculation based on only some of the variables in a data set, or to remove a particular case that is known to be an outlier. In other cases, the criteria may be based on the data; for example, you may wish to extract only those cases for which a certain variable has a particular value. In S, these kinds of tasks can usually be carried out by using expressions inside of subscripts to choose the parts of the data that you need. The first section of this chapter will deal with these sorts of tasks. The final two sections will address the problem of combining data from various objects, and specifically of rearranging the data within matrices.

5.1 Subscripting

The basic technique for accessing parts of an object in S is subscripting. Like most computer languages, S allows access through numeric subscripts, but S also allows logical and character subscripts, as well as partial subscripts, which restrict the elements to be selected in some dimensions but include all elements in other dimensions. Furthermore, special subscripting operators are available to act on lists and data frames. All these topics will be discussed in the subsections below, starting with the simplest case, that of accessing the elements of a vector.

5.1.1 Accessing Vector Elements

The methods that apply to vectors can also be extended to arrays of higher dimension, so it is worthwhile to study them carefully in this simple case. All the types of subscripting are described in the remainder of this subsection, and are summarized in Table 5.1. All of these techniques are applied to a vector

Table 5.1 Effects of Different Subscripting Techniques on a Vector

Type of subscript	Effect	Example
empty	Extract all values	`x[]`
numeric, positive	Extract the values specified by the subscript	`x[c(1,3,5)]`
numeric, negative	Extract all values except those specified by the subscript	`x[-(2:6)]`
logical	Extract those values for which the subscript is true	`x[x<7]`
character	Extract the values whose names attributes correspond to those specified in the subscript	`x["a"]`

through the use of square brackets (`[ ]`). One important feature of S is that the subscript operator is not restricted to names of data objects but can be applied to any object with multiple elements. For example, the `dim` function returns a vector containing the numbers of rows and columns of a matrix as a vector of length 2. To find out the number of columns in a matrix called `mat`, you could refer to

```
> dim(mat)[2]
```

In other words, you can apply the subscript directly to the result of the function call, without having to save the result in an intermediate vector.

Empty subscript   An empty subscript is an example of the general principle that when no value is given for a subscript, all of the elements for that dimension should be included. This will prove useful when dealing with higher-order arrays, as well as being a convenient way to preserve the length of a vector when each element is to be assigned the same scalar value. For example, if you have a vector named **x**, the assignment statement

```
> x[] <- 1
```

will assign the value of 1 to each element of **x**, preserving its length and mode, whereas the expression

```
> x <- 1
```

would change the value of **x** to a scalar whose value is 1.

Positive numeric subscript   This is the basic form of numeric subscripting, which is similar to those available in other languages. The object that serves as a subscript can be a scalar (extracting exactly one element), a vector (extracting a number of elements equal to the length of the vector), or any valid S expression that returns a numeric scalar or vector. A value of 0 (zero) as a subscript will simply be ignored. The behavior of missing subscripts (`NA`) depends on whether the missing value appears as a scalar or as an element in a vector. If a single `NA` is used as a subscript, the result will be an object of the same length as the subscripted object, with missing value for each element. If a missing value appears as an element in a vector of values used for subscripts, the corresponding element of the extracted value will be missing (**NA**). Note that the values used as numeric subscripts must be either all positive or all negative (except for zeroes and missing values), and that they need not be unique or be specified in any particular order. When a vector of subscripts is given, the extracted vector will contain the specified elements in the same order as the vector of subscripts.

Negative numeric subscript   When all the values in a numeric subscript are negative, all values except those specified in the negative subscripts are extracted. Zeroes are ignored. Numeric scalars, vectors, or expressions that return such objects may be used as subscripts in this way, provided that all the elements are negative or zero. To illustrate, consider a vector containing the powers of two:

```
> powers.of.two <- 2 ^ (1:6)
> powers.of.two
[1]  2  4  8 16 32 64
```

We could eliminate the first and last elements of the vector with the following statement:

```
> powers.of.two[-c(1,6)]
[1]  4  8 16 32
```

Logical subscript   Logical subscripts provide one of the most powerful subsetting operations available in S. When you use a logical subscript, it must be the same length as the object that is being subscripted. This is usually accomplished by using a logical expression that is operating either on the vector whose values are being extracted or on some other vector of the same length. Elements in the subscripted vector corresponding to values of **TRUE** in the subscript will be extracted; elements corresponding to values of **FALSE** will not. Missing values in the subscript result in missing values in the extracted vector. Thus, logical subscripts are a natural way to extract elements from one vector based on the value of a second vector of the same length. For example, suppose we wish to look at the values of per capita income for those states in the **state.x77** data set whose area is greater than 80,000 square miles. The expression **state.x77[,"Area"] > 80000** will result in a logical vector of length 50, with **TRUE** corresponding to those states that have areas greater than 80,000 square miles; when we use it as a subscript for income, we get the following results:

```
> state.x77[state.x77[,"Area"] > 80000,"Income"]
 Alaska Arizona California Colorado Idaho Kansas Montana
   6315    4530       5114     4884  4119   4669    4347

 Nevada New Mexico Oregon Texas Utah Wyoming
   5149       3601   4660  4188 4022    4566
```

The values of **Income** are extracted for all those states for which **Area** is greater than 80,000, because the subscripting expression **state.x77[,"Area"] > 80000** will be **TRUE** for those observations with area greater than 80,000 and **FALSE** for observations with smaller areas.

Character subscript   If a names attribute has been assigned to a vector through an assignment to the **names** function, then character values appearing in a subscript will extract the elements in the vector that have the corresponding names attribute. If there is no names attribute, or if a character string supplied as a subscript does not correspond to an element of the names attribute, a missing value (**NA**) will be extracted. Character and numeric subscripts cannot be mixed in the same expression. Also keep in mind that S is case-sensitive: if you use character subscripts, they must match exactly with respect to the cases and the punctuation in the names used.

When using a character value as a subscript, you need provide only enough characters so that S can unambiguously find the name to which you are referring. Consider a vector with two elements, named **First** and **Second**. We could access the first element as simply **"F"**, since there is no other named object in the vector that begins with the letter F:

```
> x <- c(First=1,Second=2)
> x
 First Second
    1     2
> x["F"]
 First
    1
```

If there were other named elements that began with the letter F, then you would need to provide just enough of the name so that S can unambiguously determine which element you are referring to, as the following example shows.

```
> x <- c(First=1,Second=2,Third=3,Fourth=4)
> x["F"]
 F
NA
> x["Fi"]
 First
    1
> x["Fo"]
 Fourth
    4
```

In addition to extracting elements, these subscripting operations can also be used to modify selected elements of a vector. As a simple but useful example, suppose you wish to change all values of a vector **x** that are less than zero to the value zero. If we wished to extract all the values of the vector that are less than zero, we could use the expression **x[x < 0]**. By using this expression as the left-hand side of an assignment statement, you can change all values of less than zero to zero as follows:

```
> x[x < 0] <- 0
```

When the right hand side of the assignment is a scalar value, each of the selected values will be changed to the specified value. If the right-hand side is a vector, the selected values will be replaced in order, reusing the values if more values were selected on the left-hand side than were available on the right-hand side.

As mentioned earlier, subsetting expressions are especially useful as temporary arguments to functions, because they require fewer resources than if the subsetted vector were saved as a separate object. One of the most common examples of this technique involves numeric functions that will return a missing value (NA) if any of the elements of their arguments are missing. Consider the following example:

```
> x <- c(1,5,7,12,NA,15,19)
> mean(x)
[1] NA
> mean(x[!is.na(x)])
[1] 9.833333
```

If you call the mean function directly with a vector containing one or more missing values, it will return a missing value for the mean. Since the output from is.na(x) is a logical vector of the same length as x, you can use it (or its negation) as a subscript, selecting only the elements that are not missing. Note also that in this example, you could have gotten the same result by using the expression mean(x[-5]).

A closely related problem is that of finding the locations of those elements in a vector that meet a certain condition. In a case like this, you apply the logical subscript to a vector consisting of the numbers from 1 to the length of x, instead of to x itself. The seq function is a good choice to generate this vector of indices. To find the indices of all the cars in the auto.stats data set whose turning circle is greater than 46 feet, we could use the following statements:

```
> turn <- auto.stats[,"Turning Circle"]
> loc <- seq(along=turn)[turn > 46]
> loc
[1] 29 36 37 41
```

Since loc is simply a vector of integers, you can use it (or its negative) as a subscript to extract only those values that are of interest. We could display the names of the large cars by using loc as a subscript to the first component of the auto.stats dimnames list:

```
> dimnames(auto.stats)[[1]][loc]
[1] "Dodge Magnum XE"   "Linc. Continental"
[3] "Linc. Cont. Mark V" "Merc Cougar"
```

(The double square brackets ([[ ]]) are used to access list elements; see Section 5.1.3, below, for more details.)

5.1.2   Accessing Matrix Elements

Each of the extraction methods outlined in the preceding subsection can be applied either to the first or to the second dimension of a matrix, resulting in an extremely flexible set of tools for selecting or changing elements in a matrix. The subscript in the first position will determine which rows of the matrix are to be used; the subscript in the second position will determine which columns are to be used. One caution is in order: If you accidentally refer to a matrix using only a single subscript, S will interpret this to mean that you wish to treat the matrix as a vector, with the elements of the vector corresponding to the elements of the matrix stored by columns. Consider a matrix `tst` with 10 rows and 5 columns, containing the numbers from 1 to 50 by rows. We could create such a matrix with the command

```
> tst <- matrix(1:50,nrow=10,byrow=T)
```

If you then attempted to access its elements with single subscripts, you would see the following:

```
> tst[1]
[1] 1
> tst[2]
[1] 6
> tst[3]
[1] 11
> tst[20]
[1] 47
```

Because S provides no warning messages in these cases, you should always make sure that you are accessing matrices and other arrays with the proper number of subscripts.

When a numeric vector or scalar is used as a subscript in a matrix expression, the result is a submatrix that contains some or all of the rows and/or columns of the original matrix. Considering the matrix `tst` from the example above, you can use subscripts to create a wide variety of submatrices:

```
> tst[1,3]                      # single element
[1] 3
> tst[,4]                       # column extraction
 [1]  4  9 14 19 24 29 34 39 44 49
> tst[3,]                       # row extraction
[1] 11 12 13 14 15
> tst[2:4,c(1,4,5)]             # extracting a submatrix
      [,1] [,2] [,3]
[1,]    6    9   10
[2,]   11   14   15
[3,]   16   19   20
```

```
> tst[,rev(1:dim(tst)[2])]     # rearranging columns
       [,1] [,2] [,3] [,4] [,5]
 [1,]    5    4    3    2    1
 [2,]   10    9    8    7    6
 [3,]   15   14   13   12   11
 [4,]   20   19   18   17   16
 [5,]   25   24   23   22   21
 [6,]   30   29   28   27   26
 [7,]   35   34   33   32   31
 [8,]   40   39   38   37   36
 [9,]   45   44   43   42   41
[10,]   50   49   48   47   46
```

In the previous examples, when the result of a subscript reduction was a single row or column, S automatically converted the output from a matrix to a vector. If there was a dimnames attribute associated with the matrix, then the appropriate component would be converted to the names attribute of the vector. One reason that this is done automatically is to provide a more compact representation of the data when they are printed: both rows and columns can be printed as rows. However, there are some situations in which it is necessary to make sure that the object created by the subscript reduction remains a matrix. Some S functions that expect a matrix as one of their arguments will fail when a vector is passed to them, even though it might be valid to operate on a single row or column.[1] In cases like this, you can use the optional **drop=F** argument as an additional argument to the subscript operator. Note the difference between tst1 and tst2 in the following code:

```
> tst1 <- tst[,1]
> tst2 <- tst[,1,drop=F]
> dim(tst1)
NULL
> dim(tst2)
[1] 10  1
> tst1
 [1]  1  6 11 16 21 26 31 36 41 46
```

---

[1]In the general case, a vector can always be temporarily converted to a matrix using the function as.matrix.

```
> tst2
        [,1]
 [1,]    1
 [2,]    6
 [3,]   11
 [4,]   16
 [5,]   21
 [6,]   26
 [7,]   31
 [8,]   36
 [9,]   41
[10,]   46
```

When the first column of **tst** is extracted without the **drop=F** argument (**tst1**), the result is converted to a vector. (This is indicated by the return value of **NULL** from the **dim** function.) The **drop=F** argument retains the matrix character of the result in **tst2**. This is demonstrated by the return value of **dim**, as well as by the more cumbersome printed representation. Thus, unless the **drop=F** argument is specified when a single row or column is extracted from a matrix, S makes no distinction between column vectors and row vectors; all one-dimensional collections of numbers are treated identically.

Using the **drop=F** argument also preserves any row dimnames that have been stored with a matrix, whereas a subsetting operation would leave only a single row. The data set **state.x77** included in the S distribution consists of information about the 50 U.S. states, with each state's name serving as the first component of its dimname. (For more information about this data set, use the S command **help(state)**.) The column dimname **Frost** is associated with the mean number of days per year with minimum temperatures of less than 32 degrees for each of the states. If you wish to see the variable values for states with fewer than 10 days of frost per year, you could use the following statement:

```
> state.x77[state.x77[,"Frost"] < 10,]
Population Income Illiteracy Life Exp Murder HS Grad Frost
       868    4963         1.9     73.6    6.2    61.9     0

Area
6425
```

The variable names have been correctly transferred to the resultant vector, but the row dimnames have been lost. You can ensure that they will be displayed by using the **drop=F** argument, but you must be certain that the subscripting operator knows that you wish to have an empty subscript in the second position. This explains the "extra" comma in the following statement:

```
> state.x77[state.x77[,"Frost"] < 10,,drop=F]
        Population Income Illiteracy Life Exp Murder HS Grad
Hawaii         868   4963       1.9     73.6    6.2    61.9

        Frost Area
Hawaii      0 6425
```

The frost-free state is now properly identified as Hawaii.

Logical subscripts can be used in the same fashion as numeric subscripts, keeping in mind that the length of a logical expression used as a subscript must be equal to the corresponding dimension of the matrix. For example, to extract from the saving.x data set the values of all the columns for those countries in which more than 45 percent of the population is under 15 (column 1), we could use the following expression:

```
> saving.x[saving.x[,1] > 45,]
```

This example is worth studying in some detail. The expression saving.x[,1] > 45, used as the first subscript, is a logical vector with as many elements as there are rows in the matrix saving.x, with TRUE in those positions where saving.x[,1] (the first column of saving.x, the percentage of the population under 15) is greater than 45. Because it represents the first subscript for saving.x, it will serve to select only those rows for which the first column of saving.x is greater than 45. The second subscript is empty, so all the columns of the matrix are selected.

One caution regarding the use of logical vectors for subscripting a matrix or vector concerns missing values. Since a missing value will propagate a missing value in a logical expression, the presence of missing values in your data may create unpredictable results. Suppose we wish to extract the rows of the auto.stats matrix that correspond to cars that received a repair rating of 5 in 1977. The expression

```
> auto.stats[auto.stats[,"Repair (1977)"]==5,]
              Price Miles per gallon Repair (1978)
                 NA               NA            NA
                 NA               NA            NA
Datsun 210     4589               35             5
                 NA               NA            NA
Honda Accord   5799               25             5
                 NA               NA            NA
                 NA               NA            NA
                          . . .
```

does not give the expected result. Since the expression auto.stats[,"Repair (1977)"]==5 contains a missing value in each position where the 1977 repair rating is missing, spurious observations are added to the extracted matrix. The solution is to explicitly exclude the missing values with a slightly more complex logical expression:

```
> auto.stats[auto.stats[,"Repair (1977)"] == 5 &
+              !is.na(auto.stats[,"Repair (1977)"]),]
                 Price Miles per gallon Repair (1978)
   Datsun 210    4589               35             5
 Honda Accord    5799               25             5
 Toyota Celica   5899               18             5
Toyota Corolla   3748               31             5
 Toyota Corona   5719               18             5
                        . . .
```

which correctly extracts the rows corresponding to the highly rated cars.

Beyond the techniques that are applicable to vectors, you can provide a matrix as a subscript to a matrix. As with vector subscripts, there is a major difference in the way numeric and logical matrices are interpreted when they are used as subscripts. When a numeric matrix is used as a subscript, it must have exactly two columns. The first column will serve as the row subscript; the second will serve as the column subscript. This technique provides an alternative to the subscripting methods just described, which can access only selected rectangular sections of a matrix, not an arbitrary selection of elements. When used on the right-hand side of an assignment statement, a numeric matrix subscript will result in a vector whose length is equal to the number of rows in the subscript matrix, containing the elements specified by the numbers in the matrix, treating the first column's entry as the row subscript and the second column's entry as the column subscript. For example, you could extract the superdiagonal elements of a matrix with code similar to the following:

```
> x <- matrix(1:25,5)
> x
     [,1] [,2] [,3] [,4] [,5]
[1,]    1    6   11   16   21
[2,]    2    7   12   17   22
[3,]    3    8   13   18   23
[4,]    4    9   14   19   24
[5,]    5   10   15   20   25
> supd.index <- matrix(c(1:4,2:5),ncol=2)
> supd.index
     [,1] [,2]
[1,]    1    2
[2,]    2    3
[3,]    3    4
[4,]    4    5
> supd <- x[supd.index]
> supd
[1]  6 12 18 24
```

The vector supd contains the elements x[1,2], x[2,3], x[3,4] and x[4,5].

When used on the left-hand side of an assignment statement, a numeric matrix subscript selects those elements whose subscripts correspond to the rows of the subscript matrix and replaces them one by one with the objects specified on the right-hand side of the assignment. This provides a convenient way to enter the values of a matrix from a table. For example, consider the matrix `tab1`, containing row and column indices and values for the corresponding elements of a matrix. To create a matrix containing these values in the specified locations, you could use the following S statements:

```
> tbl
      [,1] [,2] [,3]
[1,]    2    1   18
[2,]    1    1   15
[3,]    2    3   19
[4,]    1    3   12
[5,]    2    2   14
[6,]    1    2   17

> xx <- matrix(0,2,3)
> xx[tbl[,1:2]] <- tbl[,3]
> xx
      [,1] [,2] [,3]
[1,]   15   17   12
[2,]   18   14   19
```

It was necessary to create the matrix **xx** before you could refer to it in the assignment statement, so the **matrix** function was used, initially filling the matrix with zeroes. Any element of the matrix that was not specified through the subscripts in the first two columns of **tbl** would remain 0, so this technique is not restricted to the case of replacing or assigning an entire matrix with specified values.

When a logical matrix is used as a subscript, its dimensions must be identical to those of the matrix it is subscripting, and it will select those elements in the subscripted matrix for which there is a value of **TRUE** in the logical matrix. When the matrix is on the right-hand side of an assignment statement, the selected elements are placed in a vector, column by column with respect to the original matrix. As an example, we can extract all the elements less than 10 from the matrix **tst** using the statement

```
> tst[tst < 10]
[1] 1 6 2 7 3 8 4 9 5
```

We could replace all the elements less than 10 with zeroes using the statements

```
> tst[tst < 10] <- 0
> tst
        [,1] [,2] [,3] [,4] [,5]
 [1,]     0    0    0    0    0
 [2,]     0    0    0    0   10
 [3,]    11   12   13   14   15
 [4,]    16   17   18   19   20
 [5,]    21   22   23   24   25
 [6,]    26   27   28   29   30
 [7,]    31   32   33   34   35
 [8,]    36   37   38   39   40
 [9,]    41   42   43   44   45
[10,]    46   47   48   49   50
```

To restrict the assignment to a subset of a matrix, it is necessary to apply two sets of subscripts to the matrix. The first set picks out the subset of the matrix within which changes are desired; the second set creates an appropriate logical matrix defining the criteria for change. Extending the previous example, suppose we wished to convert only those elements in columns 1 and 3 of the matrix tst that were less than 10 to zeroes, leaving other elements less than 10 undisturbed. We could use the following statements:

```
> tst <- matrix(1:50,nrow=10,byrow=T)
> tst[,c(1,3)][tst[,c(1,3)] < 10] <- 0
> tst
        [,1] [,2] [,3] [,4] [,5]
 [1,]     0    2    0    4    5
 [2,]     0    7    0    9   10
 [3,]    11   12   13   14   15
 [4,]    16   17   18   19   20
 [5,]    21   22   23   24   25
 [6,]    26   27   28   29   30
 [7,]    31   32   33   34   35
 [8,]    36   37   38   39   40
 [9,]    41   42   43   44   45
[10,]    46   47   48   49   50
```

Two functions are very useful for constructing logical matrices to use as subscripts. The function row produces a matrix of the same dimensions as its argument, with values consisting of each element's row number; the function col does the same for column numbers. While these functions are not very interesting in themselves, a variety of logical expressions can be constructed from them. For example, to set every element above the diagonal of a square matrix equal to zero, we could use the following statements:

```
> tt <- matrix(1:25,5,5)
> row(tt) < col(tt)
      [,1] [,2] [,3] [,4] [,5]
[1,]    F    T    T    T    T
[2,]    F    F    T    T    T
[3,]    F    F    F    T    T
[4,]    F    F    F    F    T
[5,]    F    F    F    F    F
> tt[row(tt) < col(tt)] <- 0
> tt
      [,1] [,2] [,3] [,4] [,5]
[1,]    1    0    0    0    0
[2,]    2    7    0    0    0
[3,]    3    8   13    0    0
[4,]    4    9   14   19    0
[5,]    5   10   15   20   25
```

In a similar fashion, we could extract the diagonal elements of a matrix using the following statement (but also consider the function diag):

```
> tt[row(tt) == col(tt)]
[1]  1  7 13 19 25
```

because the row and column numbers of the diagonal elements of a matrix are equal to each other.

5.1.3 Accessing List Elements

The basic subscripting tool for lists is the dollar-sign operator ($). When components of a list are named (either through names given in a call to list or through assignment to the names function), they can be accessed by placing a dollar sign and the name of the component after the list name. For example:

```
> mylist <- list(one=c(1,2,3),two=c("a","b","c","d"),three=1:10)
> mylist$one
[1] 1 2 3
> mylist$two
[1] "a" "b" "c" "d"
> mylist$three
[1]  1  2  3  4  5  6  7  8  9 10
```

Once the components are extracted, access to the elements of the component is achieved by the methods described above, namely subscripting using square brackets. Thus, we can access the fifth element of the component named **three** in mylist as mylist$three[5].

As is the case for character subscripts (Section 5.1.1), names of list elements can be shortened, provided that S can unambiguously determine the element to which you are referring. Thus in the previous example, you could refer to elements o or tw, but not t:

```
> mylist$tw
[1] "a" "b" "c" "d"
> mylist$o
[1] 1 2 3
> mylist$t
NULL
```

When you refer to a list element that does not exist (for example, `mylist$t`), it reports a value of NULL but does not treat this as an error.

If the components of a list are not named, they can be accessed similarly to vectors, using square brackets (`[ ]`). Keep in mind that when you extract an element from an object using single square brackets, the mode of the extracted object will be the same as the mode of the original object. In particular, when single square brackets are applied to a list, the object that they will extract will also be a list, often a list containing only one object. Thus, extraction of list elements through square brackets will always result in an object of mode list, which may or may not be desirable, because many functions that expect numeric objects, for example, will not accept a list, even if it contains nothing but a numeric object. As an alternative to this means of subscripting, S also provides double square brackets (`[[ ]]`), which do not have the restriction of retaining the mode of the object on which they operate. When you use double square brackets, the mode of the extracted object will depend on what the extracted object is. For example, if a component of a list is of numeric mode, then access through double square brackets will return an object whose mode is numeric. (For vectors and matrices there is no difference between single and double square brackets; the distinction is important only for lists.) The difference between the two types of subscripts can be seen in the following statements:

```
> m1 <- mylist[1]
> m1
$one:
[1] 1 2 3
> mode(m1)
[1] "list"
> m2 <- mylist[[1]]
> m2
[1] 1 2 3
> mode(m2)
[1] "numeric"
```

The contents of `m1` and `m2` are the same, but the single brackets preserve the list mode of the extracted component, whereas the double brackets extract the object itself, a numeric vector. If the components of the list are named, a character value representing the name can be used in place of a numeric subscript. (Remember, however, that when the components of the list are named, you can extract them with the dollar-sign operator (`$`), which does not force the extracted object to be a list.) Single and double brackets work in the same fashion as when a numeric subscript is used:

```
> mylist["one"]
$one:
[1] 1 2 3
> mylist[["one"]]
[1] 1 2 3
```

As with the dollar-sign operator, once a list component has been accessed through the use of square brackets, elements of that component are referred to in the usual way. The second element of the component called **one** in **mylist** could be referred to as **mylist[["one"]][2]**.

You can also use a character vector of names inside single square brackets to extract a new list containing a subset of the named components in a list. For example, to create a list called **somelist** containing the components named **one** and **three** from **mylist**, you could use the statements

```
> somelist <- mylist[c("one","three")]
> somelist
$one:
[1] 1 2 3

$three:
[1]  1  2  3  4  5  6  7  8  9 10
```

5.1.4   Accessing Data Frame Elements

Since a data frame is conceptually a cross between a matrix and a list, it is not surprising that data frames can be accessed using any of the techniques already presented for matrices and lists. Thus, elements in a data frame can be accessed by using square brackets containing two subscripts (one or both of which may be empty) in the same way as elements of a matrix are accessed. Subsets of a data frame extracted in this way will themselves be data frames, unless they are scalars or represent a single row or column. By default, a single row extracted from a data frame will be a data frame, whereas a single column will be treated as a vector. This is a consequence of the fact that a single row of a data frame may contain mixed modes and cannot, in general, be stored in a vector. When appropriate, the defaults may be overridden by using the **drop=** argument to the subscript operator. For example, if **x.fr** is a data frame whose columns are all of the same mode, then **x.fr[1,,drop=T]** will be a vector; without the **drop=T** argument it would continue to be a data frame. If the data frame contained mixed modes, the **drop=T** argument would have no effect.

In addition to the matrix techniques, variables can be referred to by name, separated from the data frame name by a dollar sign ($), reflecting their list-like nature. A data frame also behaves more like a list than like a matrix when only a single subscript is used for extraction. Recall that when a matrix is accessed through a single subscript, it is treated as a vector with the elements stored in column-by-column order. With single subscripts, data frames behave like lists, so that a single subscript will access the corresponding column(s) (variable(s)) of the data frame. As with a list, single square brackets will access the data frame, returning an object of like mode—that is, another data frame. Double square brackets will access the actual component itself, usually a vector. The following example illustrates this behavior:

```
> xx <- data.frame(a=1:5,b=2:6,c=3:7)
> xx[1]
  a
1 1
2 2
3 3
4 4
5 5
> xx[["c"]]
[1] 3 4 5 6 7
> xx[2:3]
  b c
1 2 3
2 3 4
3 4 5
4 5 6
5 6 7
```

In the first and third cases, a data frame is extracted because single brackets were used; in the second case, the column of the data frame is extracted as a vector, because the double subscripts result in an object with the same mode as the extracted component.

An alternative means of accessing data frames is to attach the data frame to the search list, allowing the names given to variables in the data frame to be used without the name of the data frame itself. (See Section 2.4 for more information about the search list.) For example, suppose we have a data frame named kids, containing the three variables Age, Height, and Weight. The following statements would all produce a plot of Height versus Weight:

```
> plot(kids$Height,kids$Weight)
> plot(kids[,2],kids[,3])
> plot(kids[,"Height"],kids[,"Weight"])
```

An alternative would be to use the function attach to inform S that it should try to resolve names by considering the names of components of the kids data frame, as well as the objects that are stored in the search directories. After attaching the data frame, you can refer to its components directly by name, as in the following statements:

```
> attach(kids)
> plot(Height,Weight)
```

This scheme will cause no problems if the variables in the data frame are not modified. A discussion of attaching and detaching data frames when values are to be modified is presented in Section 2.9.2.

5.2 Combining Vectors, Matrices, Lists, and Data Frames

The basic tool for combining more than one S object is the function c, which is a mnemonic for *combine*. Given one or more arguments, c returns an object that contains all of the objects in the argument list, combined in such a way as to preserve the most general mode of any of the arguments. For example, when we combine a character vector and a numeric vector, the mode of the numeric vector is changed to character to allow the following concatenation:

```
> c(c("a","b","c"),c(1,2,3))
[1] "a" "b" "c" "1" "2" "3"
```

In a similar fashion, if we combine a list with a vector, the elements of the vector will be converted to a list for the following concatenation:

```
> zz <- list(1,2,3)
> xx <- c("a","b")

> c(zz,xx)
[[1]]:
[1] 1

[[2]]:
[1] 2

[[3]]:
[1] 3

[[4]]:
[1] "a"

[[5]]:
[1] "b"
```

Finally, if we combine two lists, the result is a list containing all the elements of the lists that were combined:

```
> zz <- list(1,2,3)
> xc <- list(xx)
```

```
> c(zz,xc)
[[1]]:
[1] 1

[[2]]:
[1] 2

[[3]]:
[1] 3

[[4]]:
[1] "a" "b"
```

Converting the vector **xx** to a list before the call to **c** causes **xx** to be included in the output list as a vector; otherwise, each element of **xx** would become a separate element of the output list.

Note that the modes of individual elements are preserved through the action of the **c** function. To override this preservation, and to automatically produce a vector, we can pass the optional argument **recursive=T** to the **c** function:

```
> c(zz,xx,recursive=T)
[1] "1" "2" "3" "a" "b"
```

Once again, conversion from numeric to character takes place to allow the combination to be carried out.

The use of **recursive=T** is a convenience feature, which is equivalent to explicitly breaking down each list passed to **c** before combining its arguments. The function **unlist** can be used to explicitly convert a list to a vector. The preceding example is equivalent to the following statements:

```
> c(unlist(zz),xx)
[1] "1" "2" "3" "a" "b"
```

Objects created through the use of the **c** function can be given names in the usual way—that is, through assignment to a call to the **names** function. Alternatively, if the arguments to **c** are given names when they are passed to **c**, those names will be transferred to the names attribute of the resulting object:

```
> c(a=1,b=2,c=3)
  a b c
  1 2 3
```

When dealing with matrices, the idea of combining data needs to be extended, since we can envision a matrix growing in two ways, by adding either rows or columns to the matrix. Accordingly, there are two concatenation functions available for matrices: **cbind**, for adding columns, and **rbind**, for adding rows. Either function accepts a variable number of arguments. If either of these functions is called with matrix arguments, it will check to ensure that the dimensions are appropriate; that is, if you are joining matrices by rows, they will ensure that the matrices being joined all have the same number of columns, and vice versa for joining by columns. If this is not the case, you'll see a message like this one:

```
> x <- matrix(1:50,10,5)
> cbind(x,matrix(1:20,5,4))
Error in cbind(x, matrix(1:20, 5, 4)): Number of rows of matrices
and lengths of names vectors must match (see arg 2)
```

In the example above, we were trying to join a 10×5 matrix with a 5×4 matrix; since we were binding by columns, the number of rows should have been the same. If we reshape the second matrix so that it has 10 rows, the two matrices will be compatible with respect to `cbind`:

```
> cbind(x,matrix(1:20,10,2))
       [,1] [,2] [,3] [,4] [,5] [,6] [,7]
 [1,]    1   11   21   31   41    1   11
 [2,]    2   12   22   32   42    2   12
 [3,]    3   13   23   33   43    3   13
 [4,]    4   14   24   34   44    4   14
 [5,]    5   15   25   35   45    5   15
 [6,]    6   16   26   36   46    6   16
 [7,]    7   17   27   37   47    7   17
 [8,]    8   18   28   38   48    8   18
 [9,]    9   19   29   39   49    9   19
[10,]   10   20   30   40   50   10   20
```

If any of the arguments to `rbind` or `cbind` are vectors, however, the functions will take the same approach as the `matrix` function; that is, elements of the vector will be ignored or reused to ensure conformity. This allows you, for example, to add a column of 1s to the beginning of a matrix by simply binding a 1 to the matrix in question:

```
> cbind(1,x)
 [1,] 1  1 11 21 31 41
 [2,] 1  2 12 22 32 42
 [3,] 1  3 13 23 33 43
 [4,] 1  4 14 24 34 44
 [5,] 1  5 15 25 35 45
 [6,] 1  6 16 26 36 46
 [7,] 1  7 17 27 37 47
 [8,] 1  8 18 28 38 48
 [9,] 1  9 19 29 39 49
[10,] 1 10 20 30 40 50
```

The functions will be silent about this action, unless the number of values is not an even multiple of the target size. For example, if you try to bind (by row) 3 values to a matrix with five columns, you will obtain the following warning:

```
> rbind(x,1:3)
       [,1] [,2] [,3] [,4] [,5]
 [1,]    1   11   21   31   41
 [2,]    2   12   22   32   42
 [3,]    3   13   23   33   43
 [4,]    4   14   24   34   44
 [5,]    5   15   25   35   45
 [6,]    6   16   26   36   46
 [7,]    7   17   27   37   47
 [8,]    8   18   28   38   48
 [9,]    9   19   29   39   49
[10,]   10   20   30   40   50
[11,]    1    2    3    1    2
Warning messages:
  Number of columns of result is not a multiple of
vector length (arg 2) in: rbind(x, 1:3)
```

If we temporarily coerce the vector to a matrix, `rbind` will no longer allow the concatenation:

```
> rbind(x,as.matrix(1:3))
Error in rbind(x, as.matrix(1:3)): Number of columns of
matrices and lengths of names vectors must match (see arg 2)
```

The functions `rbind` and `cbind` will also work with data frames.[2] One useful technique for adding new columns to a data frame is to call the `data.frame` function with the existing data frame as one argument, and the new column or columns as additional arguments. For example, if we had a data frame called `dframe`, and we wished to add a new column called **Num** containing the numbers from 1 to 100, we could use the following statement:

```
> dframe <- data.frame(dframe,Num = 1:100)
```

Of course, the length of a column added to a data frame must be equal to the number of rows in the data frame.

It is also possible to add rows or columns to a data frame by assigning them to a new row or column in the existing data frame. Either numeric or character subscripts are valid, and a column can also be added by assigning its value to a named component of the data frame using the dollar-sign notation. For example, we could add a new variable called **new** to a data frame with the following commands:

---

[2] In some earlier versions of S, you must coerce data frames to matrices using the function `as.matrix` in order to use `rbind` or `cbind` with data frames.

```
> xx <- data.frame(A=1:5,B=c("one","two","three","four","five"),
  C=(1:5)^2)
> xx
  A     B  C
1 1   one  1
2 2   two  4
3 3 three  9
4 4  four 16
5 5  five 25
> xx$new <- c("a","b","c","d","e")
> xx
  A     B  C new
1 1   one  1   a
2 2   two  4   b
3 3 three  9   c
4 4  four 16   d
5 5  five 25   e
```

Alternatively, the new column could have been assigned to the object xx["new"].
If a name that is already associated with a column in the data frame is used
in statements such as these, its contents will be overwritten; otherwise, a new
column will be added. To eliminate a column of a data frame, set the value of
that column to NULL. For example, to restore the xx data frame to its original
state, with three columns instead of four, use

```
> xx$new <- NULL
```

A similar effect can be achieved with numeric subscripts:

```
> xx[5] <- 1:5/10
> xx
  A     B  C new  V5
1 1   one  1   a 0.1
2 2   two  4   b 0.2
3 3 three  9   c 0.3
4 4  four 16   d 0.4
5 5  five 25   e 0.5
```

Note that the new column received a default name consisting of the letter V with
the column number appended. The new column could equivalently have been
assigned to xx[,5].

When using this technique, be sure that there are no undefined columns
between the existing data frame and the column defined by the subscript. The
same considerations apply to the case of adding new observations to a data
frame. The dim, nrow, and ncol functions can all be used to find out how many
rows and columns there are in the data frame before adding new ones.

5.3 Functions for Subsetting or Reshaping Matrices

The basic function for reshaping matrices is **matrix** itself. Recall that **matrix** is called with a vector of values, the number of rows and/or columns that the matrix is to have, and a logical argument **byrow** that informs the function as to whether it should create the matrix from the input vector column by column (the default) or row by row. A matrix can also be passed to the **matrix** function, in which case it will be treated as a vector of values in column-by-column order. Thus, we can change the shape of a matrix by passing it to the matrix function along with the new numbers of rows and columns that we desire:

```
> x <- matrix(1:30,6,5,byrow=T)
> x
      [,1] [,2] [,3] [,4] [,5]
[1,]    1    2    3    4    5
[2,]    6    7    8    9   10
[3,]   11   12   13   14   15
[4,]   16   17   18   19   20
[5,]   21   22   23   24   25
[6,]   26   27   28   29   30
> matrix(x,10,3,byrow=T)
       [,1] [,2] [,3]
[1,]     1    6   11
[2,]    16   21   26
[3,]     2    7   12
[4,]    17   22   27
[5,]     3    8   13
[6,]    18   23   28
[7,]     4    9   14
[8,]    19   24   29
[9,]     5   10   15
[10,]   20   25   30
```

Since the input matrix is converted to a vector by columns, the ordering of the elements may not be what was expected. One way to solve the problem is to take the transpose of the matrix—that is, a matrix for which the roles of rows and columns have been reversed. When the transpose of a matrix is converted to a vector, it gives a row-by-row unraveling of the original matrix, which may be more suitable for the **matrix** function. The S function that produces the transpose of a matrix is simply called **t**.[3] The following statements describe an alternative method of reshaping the matrix **x**:

---

[3]Note that t is a good candidate for the kind of confusion described in Section 2.4.

```
> t(x)
     [,1] [,2] [,3] [,4] [,5] [,6]
[1,]    1    6   11   16   21   20
[2,]    2    7   12   17   22   27
[3,]    3    8   13   18   23   28
[4,]    4    9   14   19   24   29
[5,]    5   10   15   20   25   30

> matrix(t(x),10,3,byrow=T)
      [,1] [,2] [,3]
 [1,]    1    2    3
 [2,]    4    5    6
 [3,]    7    8    9
 [4,]   10   11   12
 [5,]   13   14   15
 [6,]   16   17   18
 [7,]   19   20   21
 [8,]   22   23   24
 [9,]   25   26   27
[10,]   28   29   30
```

This kind of manipulation is especially useful if you forget to use the **byrow=T** argument of the **matrix** function when reading data with scan as described in Section 2.2.

A useful function for extracting or changing the diagonal of a matrix is **diag**. The behavior of this function varies depending on its argument, and on its placement in an assignment statement. In the simplest case, when **diag** is called with a scalar, it returns an identity matrix with as many rows and columns as its argument. (An identity matrix is a square matrix whose diagonal elements, are all 1, and whose off-diagonal elements are all 0.) When the argument to **diag** is a matrix, it returns a vector containing the diagonal elements of the matrix in question. For a square matrix, with n rows and columns, this will be a vector of length n, whereas for a nonsquare matrix, the length will be the minimum of the number of rows and columns. (For a nonsquare matrix, the diagonal is defined as those elements whose row and column subscripts are equal.) Finally, if a vector of values is assigned to a call to **diag** with a matrix argument, the diagonal elements of the matrix will be replaced by the right-hand side of the assignment. Unlike some of the other functions that substitute values, **diag** will not allow an assignment to take place unless the right-hand side is either a vector of the appropriate length or a scalar.

Exercises

1. Many statistical procedures require that a matrix be symmetric. (If the elements of a matrix A are $A_{i,j}$, the matrix is symmetric if $A_{i,j} = A_{j,i}$ for all i and j.) Using the **row**, **col**, and **all** functions, write an S expression that will test a matrix to see if it is symmetric. (*Hint:* If a matrix is symmetric, the upper triangle of the matrix will be identical to the upper triangle of the transpose of the matrix.)

2. Write an S expression to extract the row of a matrix that contains the maximum value of that matrix. What changes would be made to extract the column containing the maximum?

3. Use the subscripting operations described in this chapter to do the following:
   a. Use the **auto.stats** data set to find the largest trunk of any car that weighs less than 3000 pounds. Which car(s) have trunks of this size?
   b. Use the **saving.x** data set to find the mean savings rate for countries where the percentage of the population younger than 15 is at least 10 times the percentage of the population over 75. Also find the mean savings rate for countries where the ratio of the two percentages is less than 10. Use a boxplot or a statistical test to determine whether savings rates are different for the two groups.
   c. Use the **state.x77** data set to find the state with the minimum income in the four different regions defined by the variable **state.region**.

6 Functions in S

One of the most important features of S is the ability to write and modify functions to suit your own particular needs. Although many functions are included in the standard S distribution, there are still times when you'll need to write a function to meet a special need. Writing functions in S is easy because you write functions using the identical statements that you use when you are using S interactively. You can learn a lot about functions by examining and modifying existing functions because typing a function's name will show you the S statements that the function contains. Although the true inner workings of many S functions are hidden by calls to special functions, such as .Internal and .C, there are still many functions whose definitions are entirely written in S. Finally, when you can't find the capabilities you need anywhere in S, you can write a function that will call either C or FORTRAN code to perform your calculations and pass the results back to S.

6.1 Getting Started with Functions

The easiest way to learn about functions is to look at some existing ones. For example, consider the **matrix** function, introduced in Section 2.5.2, which converts a vector of data into a matrix. Typing the function name **matrix** at the S prompt displays the function's definition as follows:

```
> matrix
function(data = NA, nrow = 1, ncol = 1, byrow = F, dimnames = NULL)
{
        if(missing(nrow))
                nrow <- ceiling(length(data)/ncol)
        else if(missing(ncol))
                ncol <- ceiling(length(data)/nrow)
        dim <- c(nrow, ncol)
        if(length(dim) != 2)
                stop("nrow and ncol should each be of length 1")
```

113

```
            if(byrow)
                    t(array(data, dim[2:1], rev(dimnames)))
            else array(data, dim, dimnames)
    }
```

First, notice that the argument list to the function is of the form `argumentname` = `value`. This provides default values for each of the arguments to the function. Although this is not required, it is a useful technique to minimize the amount of information you need to pass to a function. (See Section 6.11.5 for more information about setting default values.) When you call a function, you can either specify all the arguments in the order in which they are found in the function definition, or name the ones you want to pass using the `argumentname` = `value` syntax. Thus, if a user fails to specify an argument called `ncol` (or if there are fewer than three arguments), S will use the default value of 1 for that argument.

A useful feature of argument matching is that S has the capability to resolve partial matches. It is not necessary to give the complete name of an argument using the `argumentname` = `value` syntax; only enough characters must be given to uniquely distinguish the argument you have in mind from the other arguments to the function. (You can access this capability directly for other purposes through the function `pmatch`.) For example, the parameter `nrow` in the `matrix` function could be abbreviated as `nr`, because there is no other argument to the function that begins with those letters. Similarly, the `byrow` argument could simply be abbreviated as `b`, because no other argument to `matrix` begins with the letter `b`.

When your function is only a single statement, there is no need to use the curly braces ({ }) after the definition of the argument list, although they can be used. For functions with more than one statement, the statements making up the function (known as the function body) must be enclosed in curly braces.

The body of the `matrix` function is straightforward: it consists of S statements to check various conditions and ascertain the values of arguments that are not specified. The function `missing` will return `TRUE` if no value for the particular argument was specified. The presence of a default value for an argument does not change the value that `missing` returns, so the first two S statements in the function will correctly determine the size of the matrix when only one of `nrow` and `ncol` is specified. Next, the number of rows and the number of columns are concatenated to form a vector suitable to be passed to the function `array`, and are checked to make sure they make sense. One of the most useful tasks that a function performs is checking that its input arguments are reasonable, so that unnecessary computations are not performed. The function `stop` can be called at any time inside a function to print an informative message, stop execution of the function, and return to the S prompt.

Finally, the `array` function is called to actually construct the matrix. With a `dim` value of length 2, `array` will form a two-dimensional matrix. The final evaluated expression inside the function body always serves as the value that the function returns. In this case there are two possibilities, depending on whether or not the **byrow=T** argument was specified. The if-else clause guarantees that only one of these expressions will actually be evaluated.

The convention of using the last evaluated statement as a function's return value usually poses no difficulties. Sometimes, however, there is no meaningful value to return, for example if a function is written exclusively for printing or plotting. In cases like this, the function **invisible**, with no arguments, can be used as the last statement of the function. This will result in a returned value of **NULL**, and no printing will take place when the function is invoked without being assigned to a target. Another use of **invisible** is to provide a mechanism whereby a value is returned from a function, but no default printing should be performed when the function is called without assigning its output to a target. In cases like this, the returned value can be passed as an argument to **invisible**.

6.2 Modifying an Existing Function Using fix

Suppose you find yourself creating many matrices from data files in which your matrix is stored row by row. It may become inconvenient to remember to specify a byrow=T argument to the **matrix** function every time you call it. One solution is to create your own function, say **my.matrix**, which will be just like the regular S function **matrix**, but will have a different default value for the parameter **byrow**. You can make a copy of the **matrix** function just by assigning it to the new name you've chosen:

```
> my.matrix <- matrix
```

When using S-PLUS, you can invoke the function **fix** to edit the **my.matrix** function and to store the edited version under the same name. (If **fix** is not available in your version of S, see Section 6.3.) The **fix** function will first open an editor window on the text of a function (based on the value of the S-PLUS system option **editor**), allow you to modify the function, and when you have completed your modifications and exited the editor, will store the modified version under its original name. Thus, the statement

```
> fix(my.matrix)
```

will open an editor on the text of **my.matrix**. You can then change the default value of the **byrow** argument using the editor in its usual fashion; when you exit the editor, the revised version will be stored in **my.matrix**. If we had not created a copy of the function, but instead edited it directly with the command

```
> fix(matrix)
```

S-PLUS would create a copy of the modified **matrix** function in your working database and print the following warning message:

```
Warning messages:
    assigning "matrix" masks an object of the same name on
    database 2
```

A message such as this should not be taken lightly! It means that whenever the function in question (in this case, `matrix`) is called in your S session, the definition that will be used is the one in your personal `.Data` directory, not the system definition. (The output of the S function `search` should show that database 2 is the system function directory.) This message will be displayed whenever you modify a function that is found in one of the directories in your search list, but not each time you call that function. As we have seen, S functions often call other S functions, so creating a local function with the same name as an S function can have unpredictable results.

6.3 Modifying an Existing Function Using Editor Commands

If the `fix` function described above is not available, you can do the same task using S editor functions directly. For example, you can use the S function `vi` to open the UNIX editor `vi` on a copy of the `matrix` function, and store the result in a new object called `my.matrix`. (You're not limited to using `vi` to edit your S objects. The function `ed` allows you to specify any editor you wish— for example, `ed(matrix, editor="jove")`. In fact, you might wish to write a function called `jove` that does exactly this.) It's important to remember that the return value from `vi` or `ed` is the modified version of the function you were editing. In particular, if you don't specify a target for one of the editor functions, it won't automatically replace your old version. If you do forget, you can retrieve your modified version by setting the function equal to the `.Last.value` variable, provided that you didn't type any more S statements after you invoked the editor. Thus, to create an edited version of `matrix`, called `my.matrix`, you could use the following command:

```
> my.matrix <- vi(matrix)
```

An editor window will be opened on the text of the `matrix` function, at which point you can edit the function as needed. Once you leave the editor, the edited version will be stored in the object `my.matrix` in your working data directory. The precautions about using names that coincide with system functions described in the Section 6.2 also apply when using the editor functions.

6.4 Errors in Writing Functions

When you modify a function and exit from the editor, S checks the syntax of the modified function to make sure that it conforms to the rules of S. If it does, the new version of the function is either stored under its original name when using `fix` or assigned to its target when using the editor functions directly. If there is an error, however, S or S-PLUS will report the type of error and the line on which it was found and will store the definition containing the error in a special temporary location. To correct the error, you need to reinvoke the editing function with no argument corresponding to the name of the object to be edited   for example, `fix()` or `vi()`. (When invoking `fix` with no argument, the most recently edited object will be displayed for editing, even if no error was made. When using the editor functions directly with no arguments, an error message will be printed if no errors have been encountered while editing objects during that session.) Once you have successfully edited the function, you can continue using `fix`, `vi`, or some other editing function as described in the preceding two sections.

6.5   Creating a New Function

Although modifying existing functions is useful for making small changes to S, there are times when you need to create an entirely new function for which there is no existing function to serve as a model. When using `fix`, you can simply provide as an argument the name of the function you wish to create; S-PLUS will produce an appropriate empty function to edit. When using an editor function directly, you will need to provide a blank template, as in the following example:

```
> newfunc <- vi(function(){})
```

In either case, once the window is opened, you can edit the function definition as you would any other file.

As an example, let's consider the common task of determining the locations of elements in a vector or matrix that meet a certain condition. We saw in Section 5.1.1 that the following simple expression will give the appropriate values:

```
seq(along=x)[cond]
```

where `cond` represents the condition of interest.

However, what happens when we attempt to use this expression on a matrix?

```
> x <- matrix(c(1,4,5,NA,3,2,NA,1,6,12),5,2)
> x
     [,1] [,2]
[1,]    1    2
[2,]    4   NA
[3,]    5    1
[4,]   NA    6
[5,]    3   12
> seq(along=x)[is.na(x)]
[1] 4 7
```

The result is a vector with the positions of the elements in the matrix when the matrix is stored column by column. A better approach would be to display the subscripts of the desired elements in a two-column matrix. One way to do this is to use the `row` and `col` functions in the following way:

```
> matrix(c(row(x)[is.na(x)],col(x)[is.na(x)]),ncol=2)
     [,1] [,2]
[1,]    4    1
[2,]    2    2
```

Our simple solution is becoming more and more complicated, especially because
we are invoking the is.na function twice with the same arguments, which clearly
is not very efficient. In situations like this, it is usually a good idea to consider
writing a function to do the job. In this example there are a number of compelling
reasons to write a function: the multiple evaluations of the is.na function, the
complexity of the matrix expression, and the fact that there are different results
for vectors and matrices. All of these issues can be dealt with once and for all
by writing a function.

The basic design of the function is as follows. There are two required arguments to the function: x, the object from which the locations will be determined,
and cond, the logical condition that will determine the locations of interest. If
x is not a matrix, we can use the simple expression based on the seq function;
if it is a matrix, we'll use the expression that uses row and col. In either case,
we must ensure that the final evaluated expression in our function contains the
answer we want, because functions in S use the value of this final expression as
their return value.

Suppose we wish to call our function where. To begin with, we can call the
function fix as follows:

```
> fix(where)
```

If fix is not available, the same effect can be achieved by calling an editor
function directly:

```
> where <- vi(function(){})
```

In either case, this will open an edit window on an empty function template.
This template can be filled with the body of the function. One approach to
writing the body of the function is as follows:

```
where <- function(x,cond){
  if(!is.matrix(x))seq(along=x)[cond]
  else matrix(c(row(x)[cond],col(x)[cond]),ncol=2)
}
```

It is a good idea to make sure the final line in your edited file is terminated by
a newline (carriage return), especially under MS-DOS or Windows.

After closing the edit window, assuming that the function entered was
syntactically correct, you will be returned to S and the function will be available
for use. We could find the locations of missing values in the matrix x with a call
to where:

```
> where(x,is.na(x))
     [,1] [,2]
[1,]    4    1
[2,]    2    2
```

6.6 Checking for Function Name Clashes

In Section 6.2, it was mentioned that functions in your working database with the same names as system functions can cause problems, especially if they are silently called by other functions. The problem was first introduced in Section 2.4 with regard to names of data objects, where the recommendation was made to change the names of offending objects to ones that do not conflict with system objects. There are several ways to check to see if functions in your working database conflict with system objects. In S-PLUS, the function **masked**, called with no arguments, will display the names of objects in your working database that are masking objects in system directories. By masking, it is meant that objects in your working database will be invoked instead of system objects of the same name. You can optionally pass either a number representing the position on the search list or a character string representing a data directory to **masked** to check a directory other than your active database. If **masked** is not available, the **match** function can be used to check for duplicate values, as implemented here in a function called **find.functions**:

```
> find.functions <- function()
+ {       sysfuns <- objects(w = 2)
|         sysfuns[match(objects(), sysfuns, nomatch = 0)]
+ }
```

Different directories can be searched by modifying the **where** argument in the two calls to **objects**. Once offending objects are discovered, they should be assigned to a different name and removed from your database.

The **conflicts** function distributed with S can also be used to find names of conflicting objects in different directories, but this function usually provides more detail than is necessary for resolving problems due to multiple names.

6.7 Returning Multiple Values

In some cases, you will find that the results of a computation cannot be conveniently returned in a single atomic object. In such cases, you can combine the returned values in a list; when the function is called it will return this list, and individual elements can be extracted in the usual way (see Section 5.1.3.) For example, suppose we wish to write a function to perform some statistical summaries for a data matrix: the number of observations, along with the calculation of the minimum, maximum, and mean for each column, as well as a covariance matrix.

We could define such a function, called **stats**, as follows:

```
> stats <- function(x){
+    mins <- apply(x,2,min)
+    maxs <- apply(x,2,max)
+    mns <- apply(x,2,mean)
+    cov <- var(x)
+    list(nobs=dim(x)[1],min=mins,max=maxs,
+         mean=mns,cov=cov)
+ }
```

(The use of the function **apply** is explained in Section 7.1.) As usual, the **stats** function will return the value of its last expression—in this case, the list containing the components of interest. The components of the list can either be stored in separate objects and combined in the call to **list**, or the expressions that calculate them can be inserted directly as arguments to the **list** function.

When the **stats** function is called, its return value could be assigned to an object, and then individual components could be accessed using the dollar-sign operator (**$**). For this reason, it is a good practice always to name the components of a list that is being returned by a function. If we invoked **stats** with **savings.x** as an argument, it would return a list containing the various statistics described above, and they could be accessed as in the following example:

```
> saving.stats <- stats(saving.x)
> saving.stats$mean
% Pop.<15 % Pop. >75 Disp. Inc. Growth Savings
   35.0898        2.293   1106.786 3.7576    9.671
```

The other components of **saving.stats** could be accessed in a similar fashion.

6.8 Errors and Warnings

One of the advantages of using functions instead of interactive code is the ability to check for potential problems in parameter values, compatibility of arguments, and so on. In well-designed functions, the bulk of the code will often be devoted to these sorts of tasks. S provides two functions that are useful in printing informative messages when a problem is discovered: **warning** and **stop**. Each of these functions prepends the function name to a message of your choice and displays the output to the standard error stream, usually the terminal. The difference between the two functions is that **warning** prints its message and then returns control to the function that has been called, whereas **stop** terminates execution of the function and returns control to the caller of the function.

As an example of the use of **stop**, consider a bivariate smoothing program that requires two vectors, **x** and **y**, which must be of the same length. The simplest use of **stop** is illustrated in the following code fragment:

```
> smoothie <- function(x,y){
+   if(length(x) != length(y))
+     stop("x and y must be of the same length")
          . . . .
+ }
```

If the function **smoothie** is called with arguments of unequal length, the following message will be printed:

```
> x <- rnorm(10)
> y <- rnorm(20)
> smoothie(x,y)
Error in smoothie(x, y): x and y must be of the same length
```

Note that you needn't concern yourself with specifying the function name in the error message; **stop** will always prepend the name and call of the function to your message.

In the above example, it might be useful to provide more information to the user, namely the lengths of the two arguments. In general, the more information you put in an error message, the more useful the error message will be. You can use the function **paste** to construct more complicated messages, including those with substituted variable values. In the above example, we could pass a call to the function **paste** as an argument to **stop**, resulting in a more informative error message:

```
> smoothie <- function(x,y){
+     if(length(x) != length(y))
+       stop(paste("lengths of x and y not equal\nlength(x)= ",
+       length(x)," length(y)= ",length(y)))
+
          . . . .
+ }
> x <- rnorm(10)
> y <- rnorm(20)
> smoothie(x,y)
Error in smoothie(x, y): lengths of x and y not equal
length(x)=  10  length(y)=  20
```

6.9 Local Variables and Evaluation Frames

In Section 2.4, the places where S looks for the objects that you refer to in an S session were discussed. Basically, objects will be found by looking in the directories defined by your search path—that is, the output of the function **search()**. When an object is created or modified, the updated version of the object will be written to the first directory on the search path. However, inside functions, this strategy could be very dangerous. Consider the case of a function that uses a temporary variable named **x** to hold an intermediate result. If the usual search rules were followed, any object stored under the name of **x** in your .Data directory would be destroyed whenever the function was called. To prevent problems like this, the usual search rules are augmented with an additional location in memory to store objects created within functions. This location is known as an evaluation frame, and its contents are known only to the function within which it was created. Note that this evaluation frame is searched before the other members of the search list, not in place of them. Thus when you refer to objects on the right-hand side of an assignment statement, if they have not been created within the function in question, they will be found by using the usual search rules. The evaluation frame also contains copies of the arguments to the functions, so modifications to objects passed to a function will only be in effect for as long as the function is active. For example, suppose you need a function that will add all the values in a vector that are greater than zero. To implement this function, you could first change all the values in the vector that are less than zero to zero, then call the **sum** function with the resulting data. (Note this is not the best way to solve the particular problem at hand.) If the function were called **sumgt0**, we could define it as follows:

```
sumgt0 <- function(x)
{   x[x<0] <- 0
    sum(x)
}
```

If a local copy of **x** were not created within the function evaluation frame, the vector **x** would be modified after the function call. However, because of the local copy of the arguments, the value of **x** in the calling environment is unchanged, as the following example illustrates:

```
> x <- c(5,6,-7,8,12,-4)
> sumgt0(x)
[1] 31
> x
[1]    5    6   -7    8   12   -4
```

Occasionally, you will want to change the value of a permanently stored object from within a function. This might be useful, for example, if a variety of quantities are being calculated within a single function, and they cannot conveniently be passed back into the calling environment as a matrix or a list. In cases such as these, you can use the special assignment operator <<-. This operator behaves exactly like the regular assignment operator <-, but it forces the assignment to be made to the working database, even when it is used inside a function.

6.10   Cleaning Up

When you are writing a complex function, it is often necessary to perform some tasks at the end of the program, such as removing temporary files or resetting options to their original values. Of course, these statements can be placed near the end of the function, but in that case they will not be carried out if the function's execution is cut short because of an error or a signal from the user. To handle such situations, you can use the S function **on.exit** to specify tasks that should be performed whenever the function terminates, whether through the usual function exit or by more extreme means. The argument to **on.exit** is an S expression to be executed before control is returned to the calling program. Note that the command is not executed when the call to **on.exit** is encountered—the expression is stored in the evaluation frame of the function, and is called immediately before the function terminates. For example, suppose we wish to write a function that will carry out a UNIX command and then call the S function **scan** to read in the results of the command.   One approach to this problem is to place the output of the UNIX command in a temporary file and then call **scan** to read this file. We can use the S function **tempfile** to generate an appropriate unique temporary filename in the directory **/tmp**. Note that **tempfile** simply returns a unique name, but doesn't create or open any files. Without any argument, **tempfile** creates files with names such as **/tmp/filennn**, where **nnn** is a number chosen to guarantee a unique name. You can pass a character string as an argument to **tempfile** to use a name other than "**file**" as a prefix. By passing a call to the S function **unlink** as an argument to **on.exit**, we can ensure that the file will be removed even if the function terminates abnormally:

```
> scancmd <- function(cmd){
+    tmpfil <- tempfile("scancmd")
+    on.exit(unlink(tmpfil))
+    .System(paste(cmd,">",tmpfil))
+    scan(tmpfil)
+ }
```

It can be seen from the preceding example that variables local to the function **tmpfil** can safely be used inside expressions passed to **on.exit** because they will be evaluated with respect to the evaluation frame of the function before it is destroyed.

6.11   Advanced Topics

6.11.1   Variable Numbers of Arguments

The functions examined in previous sections all had fixed numbers of arguments, each with a unique name. This type of function can handle a large variety of problems, but there are often situations in which there is no way to know how many arguments will be passed to a function, or when providing names for the arguments doesn't make sense. For example, suppose we wish to write a function that will return the maximum length of a series of vectors. One (unacceptable) solution would be to create a series of functions: one to handle two vectors, another for three vectors, and so on. Another possibility would

be to require the caller of the function to assemble all the vectors into a list
before calling the function, but this is clearly not a natural way to handle the
problem. S provides a mechanism for passing a variable number of possibly
unnamed arguments to a function through the use of the argument ... (three
periods). Inside the function, this collection of arguments can be converted to a
list, each element of which is one of the arguments given to the function, with
the command `list(...)`. Continuing the maximum-length example, we could
write a function as follows:

```
> maxlen <- function(...){
+    ll <- list(...)
+    mx <- 0
+    for(x in ll)mx <- max(mx,length(x))
+    mx
+ }
```

(The use of **for** loops is documented in Section 7.5.1.) We can now call `maxlen`
with as many arguments as we desire, and the function will return the maximum
length of any of its arguments:

```
> maxlen(1:10,1:20,1:40,1:30)
[1] 40
> maxlen(1:30,runif(100),1:20,1:50,1:90)
[1] 100
```

When you use ... as one of the arguments to a function, you can still have
named arguments passed to the same function, but in such a case the named
argument *must* be specified as such in the function call. In addition, the named
arguments must be placed in the argument list of the function definition after
the three periods. Continuing with the `maxlen` example, suppose we wish to
restrict the search for a maximum to only those vectors of a specific mode. We
add an optional named argument, called `mode.use`, to the definition as follows:

```
> maxlen <- function(...,mode.use="numeric"){
+    ll <- list(...)
+    mx <- 0
+    for(x in ll)if(mode(x) == mode.use)mx <- max(mx,length(x))
+    mx
+ }
> maxlen(1:20,c("A","B","C","D"),c(5,10),mode.use="character")
[1] 4
```

Unfortunately, when a variable number of arguments can be accepted by a function, S will print no warning messages if an argument name is given incorrectly.
For example, if we inadvertently called the `maxlen` function with the named argument `mode.want = "character"`, instead of the correct `mode.use`, there would
be no error or warning message, and the argument would simply be ignored. Accordingly, extra caution must be exercised when calling functions with variable
numbers of arguments. You can recognize such functions by the presence of the
argument ... either in the help file or in the function definition.

6.11.2   Retrieving Names of Arguments

It is often useful to be able to know the name of an S object that has been
passed to a function, for use in labeling of graphs or in printing error messages.
This can be accomplished by using two functions. The function **substitute** will
return an unevaluated S expression representing an object passed to it, and the
function **deparse** will take the expression and create a character representation,
suitable for printing. For example, suppose we wish to have a function that will
plot two vectors, using the names of the S objects as appropriate axis labels. We
could use **deparse** and **substitute** in the following way:

```
labplot <- function(x,y)
{  xname <- deparse(substitute(x))
   yname <- deparse(substitute(y))
   plot(x,y,xlab=xname,ylab=yname,main=
        paste("Plot of",yname,"vs.",xname))
}
```

Functions such as **exists** (see Section 2.9.3) require that their arguments
be character variables. However, it is often inconvenient to demand that the user
of a function follow this rule. For example, suppose we wish to write a function
called **which** that will tell us in which directory on the search list a particular
object will be found. We would like to be able to pass an unquoted object name
to the function, but we cannot call **exists** directly with such a name. Once
again, the function **substitute** can be used in conjunction with **deparse** to
solve the problem, as illustrated in the following function:

```
> which <- function(name)
+ { if(!is.character(name))
+    name <- deparse(substitute(name))
+    srch <- seq(along = search())
+    srch[sapply(srch,exists,name=name)]
}
```

(The use of the function **sapply** is explained in Section 7.2.) When called with
either a quoted or unquoted object name, **which** will return the locations of the
directories on the search list that contain objects with that name.

6.11.3   Operators

The importance of functions in S can be illustrated by the fact that even the
common binary operators described in Chapter 3 are actually implemented as
functions. The S language is designed so that when these operators are encountered in an S expression, they are internally converted into function calls.
For example, the multiplication operator, *, can be called as a function in the
following way:

```
> "*"(3,2)
[1] 6
```

Although this technique is definitely not recommended in practice, it illustrates that the operators in S are no different from functions. (To see the definition of an operator, you need to use the function **get** as described in Section 2.9.3, because the operator has a special meaning in S.) As mentioned previously, the operators that are part of the S language are specially recognized, so that trying to define a new operator using a single character will not be successful. However, a special type of function will always be recognized as an operator, and can be used to define new operators in the same way as new functions are constructed. This type of function begins and ends with a percent sign (%), and contains a single character between the two percent signs. Some operators of this form are included in the language. For example, the operators for matrix multiplication and integer division, described in Section 3.1, are of this form. In addition, you can create new operators of this form by using the same techniques that are used for any function.

To illustrate this technique, recall the discussion of the dot product in Section 3.6.5. The dot product of two vectors can be calculated as the sum of the products of the corresponding elements of the two vectors. We can construct an operator %.% to calculate dot products and demonstrate its use with the following statements:

```
> "%.%" <- function(x,y){
+    if(length(x) != length(y))
+        stop("lengths of x and y must be equal.")
+    sum(x * y)
+ }
> x <- 1:10
> y <- (1:10)*2
> x %.%y
[1] 770
```

This technique can be used with any single character between the two percent signs. One word of caution: Be very careful if you redefine any of the built-in S functions. If all else fails, you may need to remove them directly from the .Data directory using operating-system commands.

6.11.4 Assignment Functions

In earlier sections, functions were presented that could be used on the left-hand side of an assignment statement. For example, the function **dim** allows us to either extract or change the dimensions of a matrix:

```
> x <- matrix(1:10,5,2)
> x
      [,1] [,2]
[1,]    1    6
[2,]    2    7
[3,]    3    8
[4,]    4    9
[5,]    5   10
> dim(x)
[1] 5 2
> dim(x) <- c(2,5)
> x
      [,1] [,2] [,3] [,4] [,5]
[1,]    1    3    5    7    9
[2,]    2    4    6    8   10
```

As might be expected, the ability to use a function on the left-hand side of an assignment statement is implemented in S through a special type of function, namely one that ends with the characters <-. Functions such as these are known as assignment or replacement functions. Some examples of assignment functions that are part of the S language include levels<-, diag<-, names<-, and dimnames<-. As we've seen, you don't have to call these functions by their full names—placing the function on the left-hand side of the assignment operator (<-) tells the S interpreter to look for a function with the appropriate name. You can write an assignment function in the usual way, but you must use extra care when accessing the function because the special characters (<-) in the function's name can easily be misinterpreted. To define such a function, you must enclose its name in quotes; to edit it, you must pass a call to get to an editor function such as vi rather than refer to the function by name. In an assignment function, the final argument of the function's definition is not actually passed to the function that is called. Instead, the S interpreter automatically passes the right-hand side of the expression involving the assignment function as the last argument to that function. So, in contrast to other S functions, assignment functions are called with one less argument than is found in the function's definition.

As an example of an assignment function, consider the substring function introduced in Section 3.12. The substring function extracts part of a character string; it would be useful to construct a function that could assign a specified value to some part of a character string. A function able to do this would be called substring<-. One approach to writing such a function is to break the string into a character array containing the individual characters of the string, assigning the replacement substring to the appropriate elements of the array, and collapsing the result back into a single string. The important point to notice in writing the function is that it accepts four arguments, even though it will always be called with at most three. The S code that utilizes this approach is shown below.

```
> "substring<-" <- function(text,first,last=1000000,value){
+     if(first > last)
+         stop("must have first <= last")
+     ltext <- nchar(text)
+     last <- min(last,ltext)
+     etext <- substring(text,1:ltext,1:ltext)
+     sublen <- last - first + 1
+     etext[first:last] <- substring(value,1:sublen,1:sublen)
+     text <- paste(etext,collapse="")
+     text
+ }
```

We could use the substring assignment function to change selected parts of a string as follows:

```
> str <- "hello world"
> substring(str,7,11) <- "there"
> str
[1] "hello there"
```

If it were necessary to edit the **substring<-** function, the **get** function must be used so that the special characters in the assignment function's name do not cause confusion:

```
> "substring<-" <- vi(get("substring<-"))
```

Since the **fix** function requires an unquoted file name, it cannot be used to edit assignment functions.

6.11.5   Default Values

Earlier in this chapter, it was mentioned that default values for the arguments to a function can be set by defining the function with an argument of the form **name = value**, to set the default of the argument named **name** to the value **value**. A number of alternatives are available to give more flexibility to the design of functions. The value of the default need not be a constant; one possibility is to set a default to the result of a call to a function. For example, to ensure that a value is given for a particular argument, and that an appropriate message is printed if it is not, you can set the default for that parameter as a call to the function **stop**. Then if no value is given for the argument, a message will be printed and execution of the function will stop. For example, suppose we have a smoothing function for which we require a window **width**. We could define our function as follows:

```
> smoothie <- function(x,width=stop("A width must be specified")){
               . . .
    }
```

If the function is called without a value for **width**, a message will be printed and execution will halt:

```
> smoothie(x)
Error in smoothie(x): A width must be specified
```

Notice that S prepends the error message with an indication of the location where the error actually occurred. This is most useful when an error occurs inside a function being called by another function, and eliminates the need for the writer of the function to be concerned with this detail.

As another example, suppose we wish to print all or part of a matrix, depending on the values of an argument called **prows**. It might be convenient to have the value of **prows** default to the number of rows in the matrix passed to the function. This can easily be achieved by using the **nrow** function applied to the argument representing the matrix, as follows:

```
> printsome <- function(matrix,prows=seq(to=nrow(matrix)))
                    . . .
```

In some languages, there might be confusion as to the evaluation order of the arguments, leading to a question as to whether or not **matrix** would be properly defined when the call to **nrow** was carried out. However, S functions are governed by a principle known as lazy evaluation, which means that a default value is not evaluated until it is actually needed within the function body. The implication for setting defaults is that you can set a default value to an argument whose value may be undefined when the function is called, provided that it will be defined before the argument is actually used inside the function body. In the example above, we would certainly know what **matrix** was by the time we needed to know which rows to print, so there would not be a problem. You can rely on lazy evaluation to set defaults correctly, even if they refer to variables that don't exist at the time of the initial function call, provided that the defaults are not used in the function body before the variables in question are defined.

Another technique that can be useful when an argument will only accept one of a small number of values is to use the function **match.arg**. To use **match.arg**, set the default for the argument to a vector containing all the acceptable choices. If no value is specified for the argument in the function call, **match.arg** will return the first value specified in the default vector; if a valid value is specified it will return that value, and if an invalid value is specified it will print an appropriate error message and exit. When **match.arg** is used in this way, the name of the argument in question should be passed as the only argument to **match.arg**. For example, if we define a trivial function to exercise **match.arg**, you can see the results of all three possibilities:

```
> choice <- function(method=c("regression","anova","plot"))
    match.arg(method)
> choice()
[1] "regression"
> choice("anova")
[1] "anova"
```

```
> choice("somethingelse")
Error in call to "choice": Argument "method" should be one of:
  "regression", "anova", "plot"
```

6.12   Dynamic Loading of Outside Routines

Although the functions supplied with S allow you to perform a wide range of
tasks, there are still some things that would be difficult or awkward to do using
only the tools provided as part of the S language. In addition, iterative operations
within S are somewhat slow, because of the interpretive nature of the language.
The solution to these kinds of problems is to dynamically link subroutines or
functions written in C or FORTRAN into S, allowing you to use S's capabilities
of data management and graphics, while retaining access to the more efficient
environments that these lower-level languages provide.[1]

Two schemes are available for dynamically loading outside programs into S.
One or the other (or possibly neither or both) may be available on your system,
depending on your computer's architecture, the version of S you are using, and
your local S administrator's preferences. For simple problems, you should see no
difference between the two schemes, but for more complex problems, there are
some differences in the ways in which your compilations and loadings must be
carried out. In either case, it is important to understand how the internal S data
need to be presented to the S interface function so that they will be properly
received in the FORTRAN or C environment. This information is summarized
in Table 6.1.

Table 6.1   Types of Data in S, FORTRAN, and C

S	C	FORTRAN
"integer"	`long*`	`integer`
"single"	`float*`	`real`
"double"	`double*`	`double precision`
"character"	`char**`	`character`
"complex"	`struct {double re,im;}*`	`double complex`
"list"	`void**`	-

One point of interest to C programmers is that, regardless of the type of
data used, C functions called by S must pass only pointers to objects, never the

---

[1]If dynamic loading is not possible on your computer—for example, under MS-DOS or
Windows—a static load may provide an alternative. See the next section for information
and warnings about static loading.

objects themselves. Even if a single integer is passed to a C routine from S, it must be declared in the C routine as a pointer, and dereferenced accordingly.

It should be mentioned that debugging dynamically loaded code is somewhat awkward, especially since most symbolic debuggers cannot traverse the interface between S and any external code. For this reason it is strongly recommended that any external routines be thoroughly debugged as standalone programs before any attempt is made to dynamically load them into S. The static load procedure, described in Section 6.13, can also be used to produce a version of S that contains your external routines and may be more amenable to debugging.

The following subsection presents a simple example of dynamic loading. Then the two different methods are discussed, to allow a discussion of more complex dynamic loading problems.

6.12.1   A Simple Example

Consider the calculation of a running average for a vector of numeric values. The basic idea is to create a new vector of values in which each original value is replaced by the average of those values in the neighborhood of the original value. To control the degree of smoothing that this will create, we can vary the number of values on each side of a given value that will be included in the average. One computational nuisance is that for the first few and the last few observations in the vector, there will not be as many values to form an average as there are for values in the middle of the vector. On the other hand, if we keep a running sum of adjacent values, we need only remove one value from the sum and replace it with another as we advance through the vector. A function like this could be implemented within S, but a more efficient computation can be carried out using a routine in FORTRAN or C. We will first consider a C routine, as C routines are often easier to link with S.

C

When you incorporate an algorithm written in C into S, you must write a function with a name other than **main**. This is because the C program you write will be linked into the S executable, which is already running. Keep in mind that S has no way to access the return value of the function, however, so any communication between S and C must be carried out through arguments passed to and from the function. Extra care must be exercised because all arguments passed to the C environment must be declared as pointers, not as actual values. (This is due to the convention that in C, arguments to functions are passed by value, whereas in FORTRAN and S, they are passed by address.) This is the natural way of passing a value for a vector, but for scalars it may be necessary to modify existing code.

To clarify these points, we will implement the running-average algorithm in a C function called **runavg**. The code for this function is shown below.

```
void runavg(double *x,long n,long k,double *r)
{
    long i,j,l,u;
    double t;

    t = 0;

    for(j=0;j<=k;j++)t += x[j];

    l = 0;
    u = k;
    for(i=0;i<n;i++){
    /* the use of u - l + 1 instead of k handles the end effects */
        r[i] = t / (double)(u - l + 1);
    /* update the current running total for the next position */
        if(i > k - 1)t -= x[l++];
        if(u < n - 1)t += x[++u];
        }
}
```

This program was not specifically written to be incorporated into S, and so no care was taken to make sure that all the arguments to the function **runavg** were pointers. When you are adapting a program like this one for the S environment, it is often worthwhile to write a small function that does nothing but accept pointers from the S environment and convert them to the type of argument expected by the function containing the algorithm. The following function does just that for the running-average example:

```
void runa(double *x,long *n,long *k,double *r)
{
    void runavg(double *,long,long,double *);
    runavg(x,*n,*k,r);
}
```

Each of the arguments to the function **runa** is now a pointer, so the function **runa** is suitable for dynamic loading and execution from within S.

Assuming that both the **runa** and **runavg** C functions are contained in a file called **runa.c**, under S-PLUS the command **Splus COMPILE runa.c** will compile the program, producing an object file[2] named **runa.o**. If this facility is not available, we could compile the program using the command **cc -c runa.c**.

---

[2]The use of the word *object* in this context has no relation to the concept of an S object as used throughout the book. An object file is a file of machine instructions produced by a compiler.

(Check with your system administrator or consult your online manual if this is not the correct procedure for your computer.)

The next step in accessing an external routine from within S is to call a function that will modify the internal workings of S by including your compiled code and letting S know about the names of any new routines which you have created. Depending on your system, this function will be either **dyn.load** or **dyn.load2**. (The differences between the two functions are discussed in Section 6.12.3.) Assuming that the file **runa.c** was successfully compiled to create the object file **runa.o**, the command to dynamically load the object file is **dyn.load("runa.o")**.

Now that the code is loaded into the executing version of S, we can call it through the S function **.C**. The first argument to **.C** is the name of the subroutine that we are calling; in this example the name is **runa**, which is the same as the filename, but this need not be the case. The remaining arguments must correspond one-to-one with the arguments to the C function, and care must be taken to coerce each S object into the appropriate type as shown in Table 6.1. There are four arguments to the function: **x**, the data vector; **n**, the length of **x**; **k**, which controls the amount of smoothing; and **r**, which will hold the result of the running averages. Looking at Table 6.1, we can see that these arguments correspond to the S types of double (**x** and **r**) and integer (**n** and **k**). **.C** returns a list, each of whose elements represents the value of one of the arguments after the function has returned from the C environment. It is through this mechanism that S gets its results from your C program. As with any list, access to the list elements can be achieved through subscripting operations (see Section 5.1.3 for details). By supplying names to some or all of the arguments of **.C**, accessing elements of the returned list is made simpler, as with **z$r** in the example below. It is almost always a good idea to combine the dynamic loading and the actual call to the C routine in a function. This also provides an opportunity to set defaults, check arguments, and allocate space to hold any results that may be returned by the function. The following S function shows one possible implementation of a running average function using the C program listed above:

```
> runavg <- function(x, k = 1)
+ {    if(k <= 0)
+                stop("k must be greater than zero.")
+       dyn.load("runa.o")
+       n <- length(x)
+       r <- numeric(n)
+       z <- .C("runa",
+               as.double(x),
+               as.integer(n),
+               as.integer(k),
+               r = as.double(r))
+       z$r
+ }
```

By explicitly naming the argument **r** in the call to `.C`, we can more easily access it, since it will have that name in the list returned by `.C`. If **r** were not named, we could still access it through the expression `z[[4]]`.

When the function **runavg** is called, it will return the value of the running average as stored in the vector **r** inside the C program. Notice that before **r** was passed to the C environment, it was made a numeric vector of length **n**. If this step had not been taken, no memory would have been allocated to **r**, and the program would have failed dramatically, most likely causing the S session to terminate. This is an important point—any object passed to the C environment must be of the proper size with respect to what is expected in your C function. For arguments passed into the S function (such as **x** in this case), this will most likely already be the case. For other arguments to the `.C` function, using the **numeric** or **matrix** function is probably the easiest way to achieve this goal.

Once the function is completely debugged, you should use the `is.loaded` function to prevent unnecessary reloading. In the example above, the call to `dyn.load` would be replaced with

```
if(!is.loaded(C.symbol("runa")))dyn.load("runa.o")
```

Remember that recompiling your C program does not automatically incorporate the new version into S; to do that you must make a call to `dyn.load`. For this reason, the check using `is.loaded` should not be incorporated into your S function until you are certain that you will not be making additional changes to your C program.

FORTRAN

Most of the considerations for the C example also apply to functions written in FORTRAN. Specifically, your algorithm must be implemented as a subroutine, not as a function or main program. All communication between S and FORTRAN must be carried out through the arguments passed to and from this subroutine.

The following FORTRAN code implements the running-average algorithm:

```
subroutine runa(x,n,k,r)

double precision x(n),r(n),t
integer i,j,k,l,n,u

t = 0
do 1 j=1,k+1
1 t = t + x(j)

l = 1
u = k + 1
do 3 i=1,n
C     the use of u - l + 1 instead of k handles the end effects
r(i) = t / dble(u - l + 1)
```

```
C    update the current running total for the next position
     if(i.le.k)goto 2
     t = t - x(l)
     l = l + 1
   2 if(u.ge.n)goto 3
     u = u + 1
     t = t + x(u)
   3 continue

     return
```

Before proceeding, the FORTRAN code must be compiled into an object file. Under S-PLUS, you can compile the program using the command **Splus COMPILE runa.f**; if this facility is not available, the compilation could be done using the command **f77 -c run.f**. If this does not produce the necessary object file, check with your local systems administrator as to how to compile a FORTRAN program.

Next, the compiled code must be incorporated into S through the **dyn.load** or **dyn.load2** function. (The differences between the two functions are discussed in Section 6.12.3.) Assuming that the file **runa.f** was successfully compiled to create the object file **runa.o**, the command to dynamically load the object file is **dyn.load("runa.o")**.

Now that the code is loaded into the executing version of S, we can call it through the S function **.Fortran**. The first argument to **.Fortran** is the name of the subroutine that we are calling; in this example the name is **runa**, which is the same as the filename, but this need not be the case. The remaining arguments must correspond one-to-one with the arguments to the FORTRAN function, and care must be taken to coerce each S object into the appropriate type as shown in Table 6.1. **.Fortran** returns a list, similar to the one returned by the **.C** function in the preceding subsection, and it can be accessed in the same way.

The following function shows one possible implementation of an S function that uses the FORTRAN subroutine described above.

```
> runavg <- function(x, k = 1)
+ {
+       if(k <= 0)
+               stop("k must be greater than zero.")
+       dyn.load("runa.o")
+
+       n <- length(x)
+       r <- numeric(n)
```

```
+       z <- .Fortran("runa",
+               as.double(x),
+               as.integer(n),
+               as.integer(k),
+               r = as.double(r))
+
+       z$r
+ }
```

As in the .C example, providing a name for the final argument of the call to
.Fortran (r) allows access through the expression **z$r**. Also notice that memory
for the vector r was created through a call to the **numeric** function, so that the
FORTRAN program will have space to return its results to the S environment.

    After your program has been tested and debugged, you should check to see
if the routine you are planning to dynamically load has already been loaded in
the current S session using **is.loaded**. The preceding subsection gives details
on its use.

6.12.2   Communicating with S Functions

If you are programming in C, one additional capability that is available through
the dynamic loading facility is the ability to communicate with S functions.
Thus, functions can be defined in the S environment and then an external C
program can call these functions and receive results from them. As with the
example of dynamic loading described in Section 6.12.1, the C function that will
be accessing the S functions is not a main program or executable, but rather
a function that will be dynamically loaded into an executing version of S. An
example will illustrate the way this interface is used.

    Suppose we wish to write a program to perform numerical integration of
an arbitrary function. Although it would be more efficient to write the actual
integration program in C, the convenience of being able to write the functions
to be integrated in S cannot be denied. The key to solving this dilemma is
a function called **call_S**, which is compiled into S and will be available to an
external C routine when the external routine is dynamically loaded into S. The
algorithm which we will use to evaluate the integral is known as Simpson's rule.
(Don't let the details of the algorithm concern you; the important point is to see
how functions are accessed in the C environment.) In this method of numerical
integration, the integration interval is divided into sections, and the area of
these sections is added together to calculate the integral within the interval.
The following C function, called **simp**, performs the actual numeric integration.
It accepts a pointer to a function (within C), starting and stopping points for
the integration interval, and the number of sections to use:

```
double simp(double (*func)(),double start,double stop, long n)
{
  double mult,x,t,t1,inc;
  long i;

  inc = (stop - start) / (double)n;
  t = func(x = start);
  mult = 4.;
  for(i=1;i<n;i++)
     {x += inc;
      t += mult * func(x);
      mult = mult == 4. ? 2. : 4.; }
  t += func(stop);
  return(t * inc / 3.);
}
```

Note that no special provision is made for the use of this function with S; those details will be handled with two other C functions. The first function, named **dosimp** in the example below, is the one that will be dynamically loaded into S and called from the S environment. It extracts the S function pointer, passed as a ****void** into the C environment, and places it in a static variable that will be available to the second function, which receives arguments, passes them to S, and then receives the result through the use of the **call_S** routine. In the example below, this function is named **sfunc**. Here is the code for these two functions:

```
#include <stdio.h>
 /* declaration of static pointer to communicate to sfunc */
 static void *sfunction;

 /* function to be called from S */
dosimp(void **funclist,double *start,double *stop,
       long *n,double *answer)
{
  double sfunc(double);
  double simp(double(*)(),double,double,long);
 /* extract the internal pointer to the S routine */
  sfunction = funclist[0];
  *answer = simp(sfunc,*start,*stop,*n);
}
```

```
/* function to communicate with S */
double sfunc(double x)
{
    char *modes[1];
    char *arguments[1];
    double *result;
    long lengths[2];
    lengths[0] = (long)1;
    arguments[0] = (char*)&x;
    modes[0] = "double";
    call_S(sfunction,(long)1,arguments,modes,lengths,
            NULL,(long)1,arguments);
    result = (double*)arguments[0];
    return(*result);
}
```

The first argument to call_S in the function sfunc is the previously mentioned
S function pointer, stored in the static variable sfunction. Notice that the value
of this variable is simply passed to the C environment through the .C function
call, and then passed back to S through the call_S function. Its value should
never be modified in any way. The next argument to call_S is the number of
arguments expected by the function pointed to by sfunction. In this example,
the function to be integrated as defined in S expects only one argument. Next,
an array of the arguments themselves, cast as pointers to character, is passed to
call_S, followed by an array containing pointers to character strings containing
the modes of the argument (as described in the first column of Table 6.1) and an
array of long integers containing the length of the arguments. In this case, since
the argument to the function to be integrated is a scalar, the length is 1. The
next argument to call_S is an array of pointers to character strings containing
the names of the arguments in the S function referenced by sfunction. In the
current example, there is only one argument to the S function being called, so
no names array is necessary. If you were calling an S function omitting some
arguments, or calling the arguments in a nonstandard order, you would use the
names vector to let S know which arguments in the S function correspond to
the arguments in the arguments vector. If it is not necessary to explicitly name
the arguments, as in this case, a NULL pointer can be passed to call_S. The
last two arguments to call_S are the number of results to be passed back from
the S function, and an array of pointers to character to contain the results.
If the first of these arguments is greater than 1, the function referenced by
sfunction must return a list containing the values to be returned. Regardless
of the number of results returned, they must be cast to the appropriate type
within the C environment before they can be accessed. Note that even if you
only have one argument or result, the corresponding arguments passed to call_S
must be arrays of pointers, and not just a single pointer. If, as in this example,
there is only one argument, then the arrays are declared with a length of 1.

Assuming that the three C programs referenced above are stored in a file called `simp.c`, and have been compiled into a file called `simp.o`, an S function called `simpson` for this example is as follows:

```
> simpson <- function(thefunc, start, stop, n = 10)
+ {
+   if(trunc(n/2) - n/2!=0)
+     stop("Number of points must be even.")
+
+   if(!is.loaded(C.symbol("dosimp")))dyn.load("simp.o")
+
+   t <- 0
+   .C("dosimp",list(thefunc),as.double(start),
+             as.double(stop),as.integer(n),
+             ans = as.double(t))$ans
+}
```

The function to be integrated, `thefunc`, is passed to the C environment using the `list` function. Note that we must actually create a list to pass to C, not coerce the function to a list using `as.list`. As an example of `simpson`'s use, we can define a function within the S environment to evaluate $\cos 2x / \cos x$, and calculate the numerical integral for that function in the range of 0 to 1 with the following statements:

```
> myfunc <- function(x)cos(2*x)/cos(x)
> simpson(myfunc,0,1)
[1] 0.4567489
```

6.12.3 More Complex Issues

Missing Values

By default, the functions `.C` and `.Fortran` will not accept missing values in any of their arguments. This generally makes sense, because the internal S value for a missing value need not correspond to a missing value in the FORTRAN or C environment. If you do want to accept arguments containing missing values in your FORTRAN or C programs, however, you can pass the `NAOK=T` argument to the `.Fortran` or `.C` functions. It will then be your responsibility to correctly handle the missing values. Note that, because of partial argument matching, you should avoid using names for arguments to these functions that might be mistaken for the `NAOK` argument. Specifically, the name `N` should be avoided, although `n` will cause no problems.

Matrices

In dealing with vectors, as in the running-average example above, no problems emerge when changing between S and an outside environment, because there is only one way in which a vector can be stored in the computer. When dealing with matrices, we need to be aware of exactly how they are stored in the computer, both in S and in FORTRAN and C. As mentioned in Section 2.5.2, matrices are stored in S column by column. This corresponds directly to the way matrices are stored in FORTRAN, so in general there should be no difficulty passing a matrix from S to a FORTRAN program.[3] One problem that might arise is due to the use of the coercion functions (for example `as.double`) in the argument list to `.Fortran` or `.C`. Since these functions remove any attributes assigned to an object, a matrix passed to an outside environment in this way will not have its matrix properties when it is returned from the outside environment. In cases like this, the S function `storage.mode` can be used to alter the storage mode of the object without disturbing its matrix attributes. Suppose we have an S matrix called `mat` which we want to pass to the FORTRAN environment. Instead of calling `.Fortran` with the argument `as.double(mat)`, we could insert the following statement before the call to `.Fortran`, and pass `mat` directly to the FORTRAN environment:

```
storage.mode(mat) <- "double"
```

Alternatively, the `matrix` function could be called to convert the returned vector back to a matrix, since the internal ordering of the elements has not been changed.

When using matrices in C, attention must be paid to the way doubly dimensioned arrays are implemented in that language. In FORTRAN (and S), a matrix is stored as a contiguous collection of elements, but doubly dimensioned arrays in C do not have this constraint. In addition, row-by-row order is more natural in C than the column-by-column order used both by S and by FORTRAN. One way to cope with this problem is to write a function in C that will convert a matrix stored by columns into one that is stored by rows. To illustrate this idea, we will develop a C function that accepts a matrix and calculates the mean of the elements in each row. (This problem can be solved more simply inside S, but will serve as an example.) First, an array of pointers to double called `newmat` is allocated. Next, memory is allocated to hold the matrix, and the row pointers of `newmat` are set to the appropriate locations within this memory. Finally, the elements of the original matrix are copied into their appropriate places in `newmat`, and a pointer to `newmat` is returned. Notice that all memory allocation is performed through the function `S_alloc`. This function, which will be made available to `makemat` when the object file is dynamically or statically loaded into S, accepts two arguments, each an `unsigned` integer. The first is the number of elements desired; the second is the size of each element. (This models the standard C function `calloc`.) The advantage of using `S_alloc` instead of a memory allocation function from the C library is that memory obtained through

---

[3] Some FORTRAN subroutines require an argument representing the number of rows allocated to a matrix, or the "leading dimension" of a matrix, in addition to the actual number of rows in the matrix. When such a routine is called from S, this argument is equal to the number of rows in the matrix.

S_alloc is derived from the same evaluation frame as the function from which it is called, and will be appropriately freed when the function is exited. If you use a routine such as `calloc` to allocate memory, you are responsible for freeing any memory you allocate.

```
double **makemat(double *mat,long nrow,long ncol)
{
    long i,j;
    double **newmat,*x;
    char *S_alloc();

    newmat = (double **)S_alloc((unsigned)nrow, (unsigned)sizeof(double *));

    x = (double*)S_alloc(nrow * ncol, sizeof(double));

    for(i=0;i<nrow;i++,x += ncol)newmat[i]= x;

    for(j=0;j<ncol;j++)
        for(i=0;i<nrow;i++)newmat[i][j] = *(mat++);

    return(newmat);
}
```

With this function available, we can now write a small function to accept the appropriate (pointer) arguments from S, copy the data into the matrix, and call a C function written to accept the usual doubly dimensioned matrix.

```
void srmean(double *mat,long *nrow,long *ncol,double *means)
{
    double **mmat,**makemat();

    mmat = makemat(mat,*nrow,*ncol);
    rmean(mmat,*nrow,*ncol,means);
}

void rmean(double **mat,long nrow,long ncol,double *means)
{
    long i,j;
    double tmp;

    for(i=0;i<nrow;i++){
        tmp = 0.;
        for(j=0;j<ncol;j++)tmp += mat[i][j];
        means[i] = tmp / (double)ncol;
        }
}
```

An alternative to this scheme is to write a function that treats the matrix as a vector, remembering that the elements are stored in column-by-column order. Thus, elements in the ith row begin at the ith element of the array, and are separated by *nrow* elements. In the function that follows, the task of converting pointers from S into appropriate variables in C is carried out in the initializations of the variables nrow and ncol.

```
void vrmean(double *mat,long *nr,long *nc,double *means)
{
    long nrow = *nr, ncol = *nc;
    long i,j;
    double *mnow,tmp;

    for(i=0;i<nrow;i++){
        mnow = mat + i;
        tmp = 0.;
        for(j=0;j<ncol;j++,mnow += nrow)tmp += *mnow;
        means[i] = tmp/ (double)ncol;
        }
}
```

Assuming the function vrmean is stored in a file called vrow.c, all that remains is to compile the program and write the S function that will dynamically load the routine and pass the appropriate arguments:

```
mrow1 <- function(mat)
{
        dyn.load("vrow.o")
        res <- numeric(dim(mat)[1])
        .C("vrmean",
                as.double(mat),
                as.integer(nrow(mat)),
                as.integer(ncol(mat)),
                res = as.double(res))$res
}
```

Multiple Object Files and Libraries

It was mentioned earlier that two schemes are available for dynamic loading in S. For examples like the ones presented so far, in which all of the code to be dynamically loaded resides in a single file, there is no difference between the ways in which the two schemes are accessed. For more complicated cases, it will be necessary to determine which method of dynamic loading your version of S is using before creating the appropriate objects to be dynamically loaded.

The first scheme uses internal S code to modify the symbol table of S so that it can access your compiled program, and will only be available on machines for which the S code has been specifically modified. Its main advantage is that it can handle multiply defined names; that is, if you use a name that S is already using internally, it can still access your code without confusion. It will also usually be somewhat faster. You can recognize this method by the argument list to the function. If the only arguments to the `dyn.load` function on your system are `names` and `undefined`, then this is the method your system is using. To handle multiple object files and/or libraries under the internal method, you need to combine the pieces using the UNIX `ld` command with the `-d` and `-r` flags. (See your system's UNIX documentation for an explanation of these options.) For example, if you needed to combine the two object files `first.o` and `second.o` with the `xyz` library, you could use the UNIX command

```
ld -d -r -o all.out first.o second.o -lxyz
```

to create an UNIX object file called `all.out`. You could then dynamically load this object file into S with the command

```
> dyn.load("all.out")
```

and proceed to access your code as described in the previous subsections.

The second scheme for dynamic loading uses the incremental loading capability of the UNIX `ld` command through the `-A` flag. (Once again, check your system's UNIX documentation for a description of what `ld` is actually doing.) You can recognize this method by the presence of two additional arguments to the `dyn.load` command in S: `userlibs` and `size`. (In some versions of S, the function to perform this type of dynamic loading may be known as `dyn.load2`). A disadvantage of this technique is that no routine that you dynamically load can have a name in common with a name already in existence in the S executable. If this happens, you'll see a message such as the following (assuming a C function named "`alcvec`" was defined in the file `alcvec.o`):

```
ld: alcvec.o: _alcvec: multiply defined
ld: No such file or directory
Error in dyn.load2(names, userlibs = userlibs, s: 'ld -A ...'
      failed
```

In cases such as these, you will need to modify the offending name in your program and recompile to avoid the problem.

When using this external method, you should not combine object files and/or libraries outside of S, but rather pass them to the `dyn.load` function as the appropriate arguments. Using the previous example, we would combine the modules `first.o` and `second.o` with the `xyz` library using the following call to `dyn.load`:

```
> dyn.load(c("first.o","second.o"),userlibs="-lxyz")
```

The actual calls to the dynamically loaded code proceed as before.

Another problem that might arise when you are using this method of dynamic loading is that the object files may require more space than the default on your system. If this is the case, you'll see a message such as the following:

```
Error in dyn.load2(names, userlibs = userlibs, s:
   Whoops, we didn't ask for enough space (need 156992)
```

Simply rerun the `dyn.load2` command, using a value for the `size` argument that is slightly larger than the one printed in the error message.

6.13 Static Loading

Because dynamic loading is not available on all machines, an alternative means of incorporating C or FORTRAN programs is provided through a scheme called static loading. The basic idea behind static loading is to rebuild a version of S in the current directory that is exactly like the usual version of S, but with the addition of compiled versions of the programs you specify. This local version (which will have the filename `local.Sqpe` under UNIX or `nsqpe.exe` under MS-DOS) may be as large as two or more megabytes, so you should insure that sufficient disk space is available to hold the local copy.

When you use static loading to add a routine to S, you eliminate the need to call the `dyn.load` routine mentioned in the preceding section, because the routines you specified are compiled directly into the local copy of the executable. As an example of creating a statically loaded version of S incorporating a C program, consider the running average example introduced in Section 6.12.1. The C program to be compiled and included in the local version of S is called `runavg.c`; the command to produce the local version is `S LOAD runavg.c`. Note that this command should be made at the UNIX or MS-DOS prompt, not as a command from within S. If you call `S LOAD` repeatedly, it will store the most recent version of `local.Sqpe` in a file called `olocal.Sqpe`. To save disk space, you should remove that file as soon as you're sure that your latest version is working correctly. Under UNIX, the `S LOAD` command needs to set some default values from a makefile stored in the S home directory, so you should not have a file named `Makefile` or `makefile` in the working directory when you run the `S LOAD` command.

When you invoke S, it will check for the presence of a local version in your working directory, and, if it finds one, it will use it instead of the usual version, so you don't need to use a special command to access the executable that is produced by the `S LOAD` command. The `.C` and `.Fortran` routines are called in the same manner with static loading as dynamic loading; the only difference is that there is no need to call the `dyn.load` routine when you have incorporated your C or FORTRAN routines into S using a static load.

Exercises

1. Although it would conflict with the general philosophy of the S language, it might be useful to have a function that would accept an unquoted filename and, if no S variable with the same name existed, would scan the contents of the file. If a variable with the same name did exist, it could be used in the usual way. (Of course, since the file delimiter character (/) has special meaning within the S language, the function would be limited to accessing files within the current directory.) Use the S functions **exists**, **deparse**, and

substitute to write such a function. Do you feel that this kind of capability enhances the utility of the language?

2. Write a function to show the differences between two S functions, using the command dput to write the definitions of the functions to a file, and the UNIX command diff (or its equivalent MS-DOS command) to compare them. See Section 8.8 for information on how to communicate with the operating system to perform this task. Make sure that the command removes the files you create when it exits. *Hint:* If you are using S-PLUS, the function objdiff performs this task.

7 Programming in S

Many repetitive tasks involving matrices and vectors are carried out automatically through the use of functions and expressions in S, but sometimes more control is required in order to get a particular task done. For example, consider the problem of finding the maximum value of each column of a data matrix. A basic programming technique often used to solve such problems is to construct a loop—that is, a series of operations (function calls and/or assignments) that is repeated a number of times based on the changing value of some variable. Although S does provide the facilities to use loops in programming, it also provides some more powerful and convenient tools that operate on each row or column of a matrix, each element of a list, or each of several different groups of data based on the value of an auxiliary variable. Such functions are often referred to as mapping functions. One advantage of mapping functions over traditional programming techniques such as loops is that since S knows the size of the object in question it can determine the length of the result and the number of times the operation needs to be repeated without any further input from you. In addition, these functions are often more efficient than loop-based programs.

The first few sections of this chapter will present the mapping functions available in S. Next, other programming tools available in S will be discussed. Finally, S's implementation of object-oriented programming will be presented. Object-oriented programming is an advanced technique that allows programs to recognize their input and act accordingly. It is used in the statistical modeling functions described in later chapters, and is helpful for extending S into new fields.

7.1 Mapping Functions to a Matrix

The function `apply` will map a function to each row or column of a matrix or other array, returning the results in an appropriate object. (See the next section for information about mapping functions to vectors.) There are three required arguments to `apply`. The first is the array to be processed. The second is a scalar or vector that indicates which part of the array should be used in turn as an argument to the function to be mapped. For matrices, a value of 1 for this argument indicates that the function is to be mapped by rows; a 2 indicates that columns are to be used. Finally, the third argument is the name or definition of the function to be mapped. As a simple example, consider once again the case of finding the maximum of each column of a matrix. If we wish to find the maximum of each column of the matrix **saving.x**, we could use the following statement:

```
>  apply(saving.x,2,max)
  % Pop.<15 % Pop. >75 Disp. Inc. Growth Savings
      47.64         4.7     4001.89  16.71    21.1
```

Because the result of each application function was a scalar, **apply** combined the scalars into a vector whose length is equal to the number of columns in the original matrix. When the result of the function application is a vector, **apply** will create a matrix of values. For example, we could recode the variables in the **saving.x** data set into low and high values by applying **cut** to each column of the matrix. In this case, we need to pass an additional argument to the **cut** function, namely the number of groups to be created (2). This is accomplished by passing the additional argument to the function after the name of the function in the argument list:

```
> apply(saving.x,2,cut,2)
        % Pop.<15 % Pop. >75 Disp. Inc. Growth Savings
  [1,]       1           2          2       1       2
  [2,]       1           2          1       1       2
  [3,]       1           2          2       1       2
  [4,]       2           1          1       1       1
  [5,]       2           1          1       1       2
             .  .  .
```

If more than one additional argument were necessary, they would simply be placed at the end of the argument list. Note that **apply** demands that the first argument to the function be the piece of the array that is being operated on, so if you wish to pass that piece as some argument other than the first, you must name the additional arguments to avoid confusion. For example, if you wished to use each column of the **saving.x** data set as the y-axis of a scattergram, with the ratio of savings to disposable income as the x-axis for each of the plots, you could use the following statements:

```
  > saving.ratio <- saving.x[,5] / saving.x[,3]
  > apply(saving.x,2,plot,x=saving.ratio)
```

These ensure that **apply** will know that the columns of the matrix (because the second argument to **apply** is 2) will each successively serve as the y= argument to **plot**.

If necessary, a function definition can be substituted for a function name as the third argument to **apply**. If we wanted a vector containing the standard deviation of each column of **saving.x**, we would have to define a function to perform the calculation, since there is no built-in function to calculate the standard deviation. Because the standard deviation is defined as the square root of the variance, and the function **var** will calculate the variance, we can produce the desired result as follows:

```
> apply(saving.x,2,function(x)sqrt(var(x)))
  % Pop.<15 % Pop. >75 Disp. Inc.   Growth  Savings
   9.151523   1.290771   990.8703 2.869871 4.480407
```

Remember that the variable **x** as used in the function definition is defined locally to the function only, and will not interfere with any named object that is used elsewhere in the S session.

The same technique can be used if it is necessary to ignore missing values in an applied function. For example, if we wish to produce a table of minimum and maximum values of each column of a matrix, we can use the function **rbind** to combine the results of two calls to **apply**. Using the **auto.stats** data set, the following statements could be used:

```
> rbind(Minimum=apply(auto.stats,2,min),
+        Maximum=apply(auto.stats,2,max))
        Price Miles per gallon Repair (1978) Repair (1977)
Minimum  3291                12             NA            NA
Maximum 15906                41             NA            NA


        Headroom Rear Seat Trunk Weight Length Turning Circle
Minimum      1.5      18.5     5   1760    142             32
Maximum      5.0      37.5    23   4840    233             51


        Displacement Gear Ratio
Minimum           79       2.19
Maximum          425       3.89
```

To eliminate the missing values, we could use a simple function definition as the third argument to **apply**:

```
> rbind(Minimum=apply(auto.stats,2,function(x)min(x[!is.na(x)])),
+        Maximum=apply(auto.stats,2,function(x)max(x[!is.na(x)])))
        Price Miles per gallon Repair (1978) Repair (1977)
Minimum  3291                12              1             1
Maximum 15906                41              5             5


        Headroom Rear Seat Trunk Weight Length Turning Circle
Minimum      1.5      18.5     5   1760    142             32
Maximum      5.0      37.5    23   4840    233             51


        Displacement Gear Ratio
Minimum           79       2.19
Maximum          425       3.89
```

One additional comment about `apply` is in order. In certain cases, matrix operators will be more efficient than the use of `apply`. For example, to calculate the mean of each column of a matrix, we could sum up each column by pre-multiplying by a column of 1s, then divide by the number of rows in the matrix. In S, using the `saving.x` data set, a suitable expression would be

```
> rep(1,nrow(saving.x)) %*% saving.x / nrow(saving.x)
     % Pop.<15 % Pop. >75 Disp. Inc. Growth Savings
[1,]    35.0898      2.293   1106.786 3.7576   9.671
```

Of course, the answers are identical to those produced by the expression `apply( saving.x,2,mean)`. Although calculations that can be done directly using built-in operators will generally be more efficient than those using `apply`, the simplicity of `apply`, and the familiarity with the function as you use it, will often outweigh any loss in efficiency incurred through its use.

A problem related to simple mapping of a function is the case in which one of the arguments to the function changes as the function is mapped to each row or column of the matrix. For example, it is often useful to standardize a matrix by subtracting out the column mean from each observation. The function `sweep` is designed to achieve such a goal. As with `apply`, its first argument is the matrix that will be broken up and passed to a function, and its second argument is a scalar or vector indicating which part of the matrix or array should be passed to the function. However, the third argument is the vector of values that should be passed successively to the mapped function as a second argument. Thus, `sweep` differs from `apply` in that the function name or definition passed to `sweep` should be a function with at least two arguments, the first being the piece of the array that `sweep` will provide, and the second being the appropriate element from the vector of values passed as the third argument. Additional arguments, if necessary, are passed in the same fashion as with `apply`.

In many cases the appropriate function will be one of the binary arithmetic operators discussed in Section 3.1. In cases like these, the operator can be passed to `sweep` surrounded by double quotes (""). Suppose we wished to standardize the saving.x data set by subtracting the column mean from each observation and then dividing by the column standard deviation. (Such a standardized score is sometimes called a z-score. Variables standardized in this way have a mean of 0 and a standard deviation of 1, similarly to a standardized normal distribution.) We could produce the standardized scores using two invocations of `sweep`:

```
> mns <- apply(saving.x,2,mean)
> sds <- apply(saving.x,2,function(x)sqrt(var(x)))
> saving.z <- sweep(sweep(saving.x,2,mns,"-"),2,sds,"/")
```

```
> saving.z
```

	% Pop.<15	% Pop. >75	Disp. Inc.	Growth
Australia	-0.6271961	0.44701951	1.23416128	-0.30928223
Austria	-1.2861028	1.64010451	0.40490041	0.06007239
Belgium	-1.2336525	1.65559912	1.01091310	0.02174314
Bolivia	0.7430675	-0.48265711	-0.92611129	-1.23266881
Brazil	0.7758490	-1.13343075	-0.38180192	0.27959449
Canada	-0.3682228	0.43152490	1.89337973	-0.46259925
Chile	0.5081340	-0.73831819	-0.44801644	-0.37897179
Taiwan	1.0555838	-1.25738763	-0.82479632	0.95906762

. . .

The uses of **sweep** are not limited to the mapping of the arithmetic operators, however. Any function that accepts two arguments can be used. A common
task in data processing is converting missing value codes in a data set to the appropriate internal missing value in S, namely **NA**. Often different missing-value
codes are used for different variables in the data set. If the original data set
were called **data**, and we created a vector containing the missing value codes for
each column of the data called **misscodes**, the following statement would create
a matrix with the missing values recoded to **NA**:

```
> sweep(data,2,misscodes,function(x,miss)ifelse(x==miss,NA,x))
```

The function **ifelse** accepts as its first argument a logical vector, and returns
elements from its second argument wherever the expression is **TRUE**, elements
from its third argument wherever the expression is **FALSE**, and **NA** wherever the
expression has a missing value.

The function **outer** is a generalization of the concept of a matrix outer
product. Given two vectors a, of length m and b of length n, the outer product
of a and b is an $m \times n$ matrix, whose i,jth element is $a_i * b_j$. In a similar
fashion, **outer** takes three arguments: the first two are vectors and the third is
the function to be mapped to the vectors. If the two vectors are of length m
and n, respectively, then the output is an $m \times n$ matrix, whose i,jth element
is the result of passing the ith element of the first vector and the jth element
of the second vector to the function that is to be mapped. Note that when the
multiplication operator (*****) is the function to be mapped, **outer** is simply the
matrix outer product described above.

As an example of the use of **outer**, consider the problem of calculating
expected values under a hypothesis of no association in a contingency table.
(See Section 3.11 for a brief discussion of contingency tables.) The expected
value for a given entry in a contingency table is the product of the proportion
of observations in that row multiplied by the proportion of observations in that
column, finally multiplied by the total number of observations in the table. For
computational ease, we can use the row and column sums instead of proportions,
and just divide by the overall total. The following statements will produce a
matrix with the same dimensions as the contingency table used as input (**tabl**),
containing the expected values:

```
> overall <- sum(tabl)
> expected <- outer(apply(tabl,1,sum),
+                    apply(tabl,2,sum),"*") / overall
```

In Chapter 9 it will be seen that **outer** is also useful for generating function values in the course of producing contour and 3-D graphics.

7.2 Mapping Functions to Vectors and Lists

The functions in the previous section allow mapping either to rows or to columns of a matrix, but will not be suitable for objects without the regular structure of a matrix, such as vectors or lists. Two other mapping functions are available for these purposes: **lapply** and **sapply**. Each of these functions operates on either lists or vectors; **lapply** returns its results in a list, with one element in the output list for each element in the input vector or list, whereas **sapply** returns its output in the form or a vector or matrix if the output of the mapped function will allow it.

One important use of these functions is for working with data frames that have both numeric and character variables, since such objects cannot be coerced into matrices. (Note that if all the columns of a data frame are numeric, the techniques described in the preceding section can be used without modification.) To produce vectors containing column-by-column summaries similar to those described in the preceding section for matrices, it is sometimes necessary to treat nonnumeric columns differently from numeric ones. Suppose we create a data frame from the **auto.stats** matrix by adding a column for the **cargrp** variable defined in Section 3.11. We could create a vector containing the maximum value for each numeric variable in the data frame, with a missing value for nonnumeric variables, using the following expression:

```
>   cargrp <- cut(auto.stats[, 8], c(0, 2500, 3500, 5000),
+                 labels = c("Small","Medium","Large"))
> auto <- data.frame(auto.stats,Group=cargrp)
> sapply(auto,function(x)if(is.numeric(x))max(x[!is.na(x)]) else NA)
  Price Miles.per.gallon Repair..1978. Repair..1977. Headroom
  15906               41             5             5        5

Rear.Seat Trunk Weight Length Turning.Circle Displacement
     37.5    23   4840    233             51          425

Gear.Ratio Group
      3.89    NA
```

As another example of working with data frames, we can use **sapply** as the core of a function which will extract only the numeric columns of a data frame:

```
> num.frame <- function(frame)frame[,sapply(frame,is.numeric)]
```

This function will accept a data frame as an argument, and will return another data frame containing only the numeric variables in the original.

Finally, `lapply` or `sapply` can be used with data frames in conjunction with the function `split` (Section 3.11). When applied to a data frame, `split` will create a list of data frames with the same variables as the original data frame; each of these data frames will contain those observations from the original data frame with different values for a classification variable that is supplied as the second argument to `split`. For example, suppose we wished to divide the states in the `state.x77` data set into three groups, based on the rate of illiteracy in each state, and then to produce a set of scatterplots for each group using the `pairs` function. The following statements would achieve this goal:

```
> state <- data.frame(state.x77)
> lapply(split(state,cut(state$Illiteracy,3)),pairs)
```

The function `tapply`, described in the next section, performs a similar function on a vector without the need for a call to `split`.

By using a function that returns a logical vector, `sapply` can be useful in many types of subsetting operations. For example, suppose we wish to extract data for those cars from the `auto.stats` data set whose engines have displacements of 97, 134, or 231 cubic inches. We could construct a logical expression directly, but we would need to explicitly test for equality to each of the three values. An alternative is to use the `any` function, applied to each element of the `auto.stats[,"Displacement"]` vector as follows:

```
> auto.stats[sapply(auto.stats[,"Displacement"],
+                   function(x)any(x==c(97,134,231))),]
```

	Price	Miles per gallon	Repair (1978)
Audi Fox	6295	23	3
Buick Le Sabre	5788	18	3
Buick Riviera	10372	16	3
Buick Skylark	4082	19	3
Olds Cutlass	4733	19	3
Olds Cutl. Supr.	5172	19	3
Olds Delta 88	5890	18	4
Olds Omega	4181	19	3
Pont Catalina	5798	18	1

. . .

Note the use of commas as appropriate place holders, to make sure the entire row of the data set is extracted.

7.3 Mapping Functions Based on Groups

The previous sections have all dealt with cases in which the organization of the data was dictated by the way it was stored—either as a matrix organized into rows or columns, or as a list or vector, broken down into separate elements.

Often it is necessary to apply a function to groups of data based on other criteria. For example, we might have data on the locations of a collection of plants, and need to know the mean value of a variable measured on those plants, broken down by the locations from which the plants were collected. There may be several grouping variables that are of interest. If we took a survey of voters, we might want to perform analyses or display data separately for groups defined by political affiliation and level of income. The function `tapply` can be used to do this, by mapping a function to different groups of data based on the values of one or more grouping variables. The first argument to `tapply` is the vector of data to be broken down by groups, the second argument is a vector or list containing the grouping values, and the third argument, which is optional, is the function to be applied to each part of the data in turn. As a simple example, consider the `auto.stats` data set. Suppose we wish to calculate the mean gas consumption (column 2) for the three groups of cars (Small, Medium, Large) defined in Section 3.11. The following expression will map the `mean` function appropriately:

```
> tapply(auto.stats[,2],cargrp,mean)
     Small    Medium Large
  27.52174 20.12903 15.95
```

Since the `cargrp` variable was defined as a factor, with appropriate labels, `tapply` uses those labels as the names attribute of the resulting vector.

If we want to break down a vector by more than one categorical variable, a list containing the variables that define the multiple categories is used as the second argument to `tapply`. To extend the previous example, suppose we wished to view the mean gas consumption for different-sized cars, as described by the `cargrp` variable, broken down by the 1977 repair rating (column 4). The following statements will produce a matrix with the desired means:

```
> tapply(auto.stats[,2],list(cargrp,auto.stats[,4]),
+          function(x)mean(x[!is.na(x)]))
            1    2        3        4     5
  Small 21.0   NA 25.25000 31.71429 27.25
 Medium 20.5 20.5 19.85714 21.66667 18.00
  Large   NA 16.0 15.80000 16.00000    NA
```

Missing values have been placed in the output for those cross-classifications that had no observations. (The function `table`, described in Section 3.11, could be used to find the number of observations in each cell.)

To use `tapply` on a number of variables at once, it is often useful to use `apply` to map `tapply` to each of the variables in turn. For example, suppose we wish to calculate the mean of columns 5 through 8 of the `auto.stats` data set broken down by the `cargrp` variable of the previous example. The following statements produce a matrix containing the desired means:

```
> auto.means <- apply(auto.stats[,5:8],2,
+                       function(x)tapply(x,cargrp,mean))
```

```
> auto.means
          Headroom Rear Seat    Trunk   Weight
  Small 2.565217   24.86957 10.95652 2090.435
 Medium 2.870968   26.64516 12.80645 3060.968
  Large 3.650000   29.32500 18.40000 3991.500
```

Note that the levels labels and variable names were automatically carried through the computation by the mapping functions, producing an easy-to-understand display. An alternative technique using row indices is presented later in this section. It allows access to more than one variable at a time in the mapped function.

When **tapply** is called without a third argument, it returns a vector of the same length as its first argument, containing an index into the output that normally would be produced by **tapply**. This vector can then be used to identify the output from the mapped function that corresponds to each observation in the data. In the previous example, we could subtract the cell mean defined by the cross-classification from each observation using the statements shown below. The final statement combines the classification variables with the resultant vector for a more convenient display. The call to **data.frame** is used instead of **cbind** because it allows the printing of character values for the **cargrp** variable.

```
> where <- tapply(auto.stats[,2],list(cargrp,auto.stats[,4]))
> means <- tapply(auto.stats[,2],list(cargrp,auto.stats[,4]),
+                 function(x)mean(x[!is.na(x)]))
> newgas <- auto.stats[,2] - means[where]
> data.frame(Size=cargrp,Rating=auto.stats[,4],Deviation=newgas)
                 Size Rating  Deviation
  Amc Concord Medium     2   1.5000000
    Amc Pacer Medium     1  -3.5000000
   Amc Spirit Medium    NA          NA
   Audi 5000 Medium     2  -3.5000000
    Audi Fox  Small     3  -2.2500000
   BMW 320i Medium     4   3.3333333
Buick Century Medium    3   0.1428571
```

Thus, considering cars of the same size and with the same repair rating, the Amc Pacer has lower gas consumption than the average, whereas the BMW 320i has greater gas consumption than the average.

If the grouping index is to be used to access the rows of a matrix or a data frame, you can pass a vector containing a sequence of numbers from 1 to the number of rows in the matrix as the first argument to **tapply**, and then use it as a subsetting subscript in the function definition passed to **tapply**. To perform a linear regression on gas consumption as a function of weight for each of the three car sizes in the **auto.stats** data set, we could use the expressions given below. Since the output from the **lm** command cannot be put in a matrix or a vector, it is returned as a list, and is displayed by mapping the function **coef** to each element of the list using **lapply**:

```
> regrs <- tapply(seq(len=dim(auto.stats)[1]),cargrp,
+                   function(i)lm(auto.stats[i,"Miles per gallon"] ~
+                                 auto.stats[i,"Weight"]))
> lapply(regrs,coef)
$Small:
 (Intercept) auto.stats[i, "Weight"]
    51.04225               -0.01125149

$Medium:
 (Intercept) auto.stats[i, "Weight"]
    37.63337               -0.005718562

$Large:
 (Intercept) auto.stats[i, "Weight"]
    29.89971               -0.003494855
```

7.4 Conditional Computations

The basic tool for conditional computations is the **if** statement. The form of
the **if** statement is

```
if(condition)expression-1
else expression-2
```

where the **else** statement and the accompanying expression are optional. The
terms *expression-1* and *expression-2* are either single S expressions or collections of one or more S expressions surrounded by curly braces ({ }), and
condition is an S expression that is evaluated as either **TRUE** or **FALSE**. In addition to the **if** statement, S also provides an **ifelse** function, described below.
Several **if** statements can be nested within each other to test successively for
more than one condition. Once a condition is encountered that returns the value
FALSE, the remaining conditions are not actually evaluated. Thus a statement
such as

```
> if(x > 0)if(log(x) > 2)y <- log(x)
```

will ensure that the logarithm of **x** will not be evaluated unless **x** is greater than
zero. Also see Section 3.2 for a discussion of the logical operator **&&**, which is
useful in situations like these.

It is also possible to test for a variety of different possibilities by combining
several **if-else** statements. For example, suppose we are writing a function
that will perform different analyses based on the value of a character variable
called **code**. We could test the value of **code** and perform the appropriate action
with code such as the following:

```
if(code == "plot")plot(x,y)
else if(code == "regr")lm(y ~ x)
else if(code == "robust") rreg(x,y)
else print(paste("Don't understand code ",code))
```

This construction is slightly more efficient than an unconnected series of `if` statements, because subsequent `else` clauses need not be evaluated once a true condition is found. Also see the description of the `switch` function later in this section for an alternative means of choosing among a variety of possibilities for conditional computing.

The `ifelse` function accepts as its first argument a logical vector or matrix (or an expression that results in a logical vector or matrix). The object returned by `ifelse` is of the same length and dimensions as its first argument, and contains the corresponding element of the second argument wherever the first argument was `TRUE`, the corresponding element of the third argument wherever the first argument was `FALSE`, and `NA` wherever an `NA` was encountered in the first argument. If the dimensions of either the second or the third argument are not the same as those of the first argument, their values are silently recycled. In particular, either of the second and third arguments may be a scalar. Note that the return value from `ifelse` is always an object of the same length and dimensions as its first argument. For example, we could recode a matrix of values as 0 if the value was less than 10 and 1 if it was greater than 10, with a statement such as the following:

```
> x <- matrix(1:25,ncol=5)
> x
     [,1] [,2] [,3] [,4] [,5]
[1,]    1    6   11   16   21
[2,]    2    7   12   17   22
[3,]    3    8   13   18   23
[4,]    4    9   14   19   24
[5,]    5   10   15   20   25
> ifelse(x < 10,0,1)
     [,1] [,2] [,3] [,4] [,5]
[1,]    0    0    1    1    1
[2,]    0    0    1    1    1
[3,]    0    0    1    1    1
[4,]    0    0    1    1    1
[5,]    0    1    1    1    1
```

The `ifelse` function will always evaluate both its second and third arguments, so it may produce spurious warning or error messages. For example, suppose we wish to calculate the logarithm of a vector of values provided the value is greater than zero, and to substitute zero if that condition is not met.

```
> x<-c(2,3,-4,5,-6)
> ifelse(x > 0,log(x),0)
[1] 0.6931472 1.0986123 0.0000000 1.6094379 0.0000000
Warning messages:
  NAs generated in: log(x)
```

The warning message appears even though none of the NAs generated was actually used in producing the result.

For the closely related task of recoding values, a set of if-else statements similar to the ones presented above could be used, but another useful alternative is the match function. This function accepts as its first argument a vector of values (x), and as its second argument a vector of the possible values that may be found in the first argument (table). It returns a vector, the same length as x, containing the indices of the values of x in table. This returned value can be used as a subsetting subscript for the vector of recoded values. For example, suppose we wished to rescale the 1978 automobile ratings of the data set auto.stats, so that values of 1 through 5 were recoded as 1, 2, 5, 10, and 20, respectively. The following statements would achieve this goal, using cbind to display the old and new ratings side by side:

```
> old.vals <- 1:5
> new.vals <- c(1,2,5,10,20)
> new.rating <- new.vals[match(auto.stats[,"Repair (1978)"],old.vals)]
> cbind(old.rating=auto.stats[,"Repair (1978)"],new.rating)
```

	old.rating	new.rating
Amc Concord	3	5
Amc Pacer	3	5
Amc Spirit	NA	NA
Audi 5000	5	20
Audi Fox	3	5
BMW 320i	4	10
Buick Century	3	5

. . .

By default, match returns a missing value whenever a value in x is not in table. In the above example, this causes missing values to be propagated in the output. If a different value for nonmatches is desired, it can be specified with the optional nomatch= argument.

As mentioned previously, an alternative to a series of if-else statements is available through the switch function. The first argument to switch is an S expression that is evaluated as either a character or an integer value. If it is a character value, then the remaining arguments to switch are searched for one whose name corresponds to the expression's value; the value with that name is then returned. By placing a function call as the named argument, switch can provide an alternative to the if-else example presented above:

```
> switch(code,plot=plot(x,y),regr=lm(y ~ x),robust=rreg(x,y))
```

If the value of code does not match any of the available choices (plot, regr, or robust), switch will return a value of NULL.

If the first argument to switch is a numeric value, it is first truncated to an integer if it is not already one, and, considering the remaining arguments to the function, switch returns the ordinal argument corresponding to the value of this integer, provided that enough arguments are passed to the function; otherwise it returns a value of NULL. Unlike the match function, switch will work correctly only with a scalar as its first argument. To carry out the recoding example for a single value (not a vector) using switch, we could use the following statements:

```
> value <- 3
> switch(value,1,2,5,10,20)
[1] 5
```

Since value was equal to 3, switch returns the third argument after value in the argument list, namely 5.

7.5 Loops

7.5.1 for Loops

The for loop is the basic looping construct in S. The form of the for loop is as follows:

```
for(name in values) expression
```

The action of the for loop is to replace the value of name with each element in values in turn, and then to evaluate expression. If there is more than one S statement in expression, the multiple statements should be surrounded by curly braces ({ }); a single statement can be typed either on the same line as the for statement or on a separate line. The object referenced by name is local to the statements inside the loop; if there is an object with the same name stored in a data directory, its value will not be changed. One important consideration about for loops (and, in fact, about loops in S in general) is that the usual default of displaying the value of an expression that is not assigned to a value is *not* in effect within expressions controlled by a loop. This means that if you wish to see the results of calculations performed within a loop, you must explicitly display them, using the print function or the techniques described in Chapter 8.

In the above definition, values can be either a vector or a list, allowing great flexibility in the action of the loop. For example, in Section 7.2, we used the split function to create a list containing three data frames, depending on the rates of illiteracy in the states. We could perform an analysis for each of these data frames with statements such as the following:

```
> state <- data.frame(state.x77)
> for (data in split(state,cut(state$Illiteracy,3))){
statements for analysis of data
}
```

Within the statements in the body of the loop, data would be equal to each of the three data frames in the list in turn.

When you are using this technique, it is sometimes useful to have access to the names of the list elements as they are being accessed in the loop. For example, the cargrp variable created in Section 3.11 had descriptive labels (Small, Medium, and Large), which might be useful for labeling output. In cases like these, you can loop over the names of the data frames and access the elements using these names and double square brackets. As an example, suppose we wish to calculate the mean gas consumption (Miles per gallon) for cars in each of the three categories defined by cargrp, and print a sentence presenting this information. By using a loop based on the names of the list elements produced by split, we can produce the desired descriptive labels:

```
> auto.split <- split(as.data.frame(auto.stats),cargrp)
> for(nm in names(auto.split)){
+   auto.now <- auto.split[[nm]]
+   cat(dim(auto.now)[1],"cars in the",nm,
+       "group, with a mean consumption of",
+       format(round(mean(auto.now$Miles.per.gallon),2)),
+       "MPG.\n")
+ }
23 cars in the Small group, with a mean consumption of 27.52  MPG.
31 cars in the Medium group, with a mean consumption of 20.13  MPG.
20 cars in the Large group, with a mean consumption of 15.95  MPG.
```

Loops are often used to collect together the results of repetitive tasks such as simulations into a vector or matrix. In cases like this, you can either add to the result at each iteration of the loop, using c, cbind, or rbind, or create an object of appropriate size before the loop and refer to its elements on the left-hand side of an assignment statement. For example, suppose we wish to perform a simulation wherein we generate 1000 sets of two vectors of 100 random numbers from the Gamma distribution (with shape parameter equal to 1) and calculate the slope and intercept of the linear regression between them. Since we know that we will have 1000 sets of two numbers in the result, we can conveniently store the results in a matrix, and use the matrix function to prepare a matrix to receive the results before the loop itself:

```
> result <- matrix(0,1000,2)
> for(i in 1:1000){
+       x <- rgamma(100,1)
+       y <- rgamma(100,1)
+       result[i,] <- coef(lm(y~x))
+       }
```

The first column of the **result** matrix would contain the values estimated for the intercept, and the second column would represent the slope. An alternative means of creating the **result** matrix is to use **rbind** to add each set of coefficients to the matrix as they are calculated. In cases such as these, you can assign a value of **NULL** to the object before the loop:

```
> result <- NULL
> for(i in 1:1000){
+       x <- rgamma(100,1)
+       y <- rgamma(100,1)
+       result <- rbind(result,coef(lm(y~x)))
+       }
```

In general, if you know the size of the output object before beginning the loop, it is more efficient to access it directly rather than relying on **rbind**, **cbind**, or **c**.

A **for** loop can be useful in selecting rows or columns of a data set that meet certain criteria. For example, suppose we wish to remove from a matrix all rows that contain any values less than zero. The following loop will generate a logical vector called **omit**, which will contain **TRUE** for any row in a matrix **x** that has elements less than zero:

```
> omit <- F
> for(i in 1:dim(x)[2])omit <- omit | (x[,i] < 0)
```

This vector could then be used in an expression such as **x[!omit,]** to select only those rows that contain values less than zero. Note that after the first iteration, **omit** is changed from a scalar to a vector whose length is equal to the number of rows in the matrix **x**.

It should be reiterated that many operations that might be handled by loops in other languages can be more efficiently performed in S by using the subscripting facilities described in Section 5.1.1. As an example, consider the task of changing all values in a matrix that are less than zero to a value of zero. Using loops, we could write:

```
> for(i in 1:dim(x)[1])
+    for(j in 1:dim(x)[2])
+       if(x[i,j] < 0)x[i,j] = 0
```

However, the identical result can be achieved by using the following simpler (and much more efficient) statement:

```
> x[x<0] <- 0
```

The same technique can be used whenever you are looping through each element of a matrix or a vector and changing a value based on a logical expression. See Section 5.1.1 for more information about this capability.

7.5.2 while Loops

The **for** loop is useful when you know in advance how many times you will need to perform the statements in the loop, as will usually be the case when you are operating on vectors, matrices, or lists. However, there are times when other criteria must be used to determine how many times the statements in a loop are carried out. In S, problems like this can be handled with a `while` loop. The basic structure of the `while` loop is

```
while(condition) expression
```

As in the `for` loop, *expression* can be either a single S statement or a group of S statements surrounded by curly braces. The *condition* is evaluated, and, if it is `TRUE`, the *expression* is evaluated. This process continues until the *condition* is `FALSE` or the loop is interrupted. (See the next section.)

A wide variety of problems can be effectively solved by iterative methods— that is, techniques in which calculations are carried out until an answer converges to a stable value. This condition is usually detected by comparing the current answer to the answer calculated in the previous iteration, not by counting the number of times the statements in the loop have been executed. For example, Newton's method of finding the roots (zeroes) of a function uses the following formula to update its estimate at each iteration:

$$x_{i+1} \;=\; x_i \;-\; \frac{f(x_i)}{f'(x_i)}$$

where $f(x)$ is the function whose roots are desired and $f'(x)$ is the derivative of that function. The iterative process is stopped when the absolute value of the difference between the new value (x_{i+1}) and the last value (x_i) becomes sufficiently small. We could find a real root of the polynomial $x^3 - 2x^2 + 3x - 5$ starting with an initial guess of 1, using the following code:

```
> f1 <- function(x){x^3 - 2 * x^2 + 3 * x - 5}
> f2 <- function(x){3 * x^2 - 4 * x + 3}
> lastx <- 1000
> x <- 1
> while(abs(x - lastx) > 1.e-8){
+        lastx <- x
+        x <- x - f1(x) / f2(x)
+        }
> x
[1] 1.843734
> f1(x)
[1] -8.881784e-16
```

7.5.3   Control Inside Loops: next and break

When calculations are carried out inside loops, sometimes a condition occurs such that the remaining statements inside the loop need not be carried out for the current iteration. In cases such as these, the `next` command can be used to instruct S to skip over any remaining statements in the loop and continue executing. In the case of a **for** loop, the loop variable is also advanced to its next value before execution at the beginning of the loop is begun. When several loops are nested within each other, the **next** statement applies to the most recently opened loop.

Whereas the **next** command simply skips over the remaining statements inside a loop but continues executing, the **break** statement causes S to immediately stop calculations and exit from the loop. In interactive sessions, the usual prompt will reappear and you can continue typing statements; in a source file or function, execution will continue with the next statement encountered after the loop. As with the **next** statement, a **break** statement inside nested loops applies only to the most recently opened loop.

7.5.4   repeat Loops

The **while** loop is useful when there is a single, clearly defined criterion for determining when a loop should stop executing, as in the preceding example using Newton's method. In some cases there may be more than one criterion, and these may be calculated at various points inside the loop. In cases such as these, it may be more convenient to use a **repeat** loop. The syntax of the **repeat** loop is particularly simple:

```
repeat expression
```

where **expression** is either a single statement or a series of statements surrounded by curly braces. The **expression** will be evaluated continuously, so a **break** statement should always be included inside the loop to guarantee that it will eventually stop executing.

For example, suppose we wish to calculate the differences between the mean values of a time series before and after each point in the series, to determine if the value ever exceeds some limit. We would stop the calculations either if a large enough difference were found, or if all the remaining values in the series were missing. If the time series is named **x**, and the limit's value is **lim**, we can use the following **repeat** loop to carry out the calculation:

```
> i <- 1
> lenx <- length(x)
> repeat{
+   if(i > lenx)break;
+   if(all(is.na(x[i:lenx])))break;
+   difference <- mean(x[1:i]) - mean(x[(i + 1):lenx])
+   if(abs(difference) > lim)break;
+   i <- i + 1
+   }
> if(i > lenx)cat("Limit not exceeded\n") else
+     cat("Limit found at position",i,"\n")
```

In many cases, the choice between a `while` loop and a `repeat` loop is simply a matter of preference.

7.6 Advanced Topics

7.6.1 Calling Functions with Lists of Arguments

Since a wide variety of functions in S accept variable numbers of arguments, it is often useful to be able to call an S function without explicitly building the argument list and calling the function directly. The function `do.call` allows you to specify the name of a function and provide the arguments to the function in the form of a list, as an alternative to the usual form of a function call. Named arguments in the list passed to `do.call` can be used to pass specific arguments to the function being called. Since data frames have all the properties of lists, `do.call` often provides a convenient way of dealing with all of the columns (variables) of a data frame in a function call. To illustrate the use of `do.call`, consider a function to detect patterns of missing values in a data frame. The basic strategy of the function is to call the function `is.na` to produce a matrix of `TRUE` and `FALSE` values, convert those values to 0s and 1s, paste together each row of the matrix to form a character value representing a pattern of missing values, and finally call the function `table` to count the number of times each pattern is encountered. Since we want the function to be able to handle any number of columns in its input data frame, the call to `paste` will be carried out using `do.call`. Rather than using the coercion function `as.integer` to convert logical values to numbers, the storage mode of the logical matrix is changed to integer, to preserve the matrix nature of the object:

```
> na.pattern <- function(frame){
+       nas <- is.na(frame)
+       storage.mode(nas) <- "integer"
+       table(do.call("paste",c(as.data.frame(nas),sep="")))
+ }
```

The c function is used to combine the columns of `nas` with the named argument (`sep`) to `paste`. Applying `na.pattern` to a data frame created from the `auto.stats` matrix yields the following:

```
> na.pattern(as.data.frame(auto.stats))
 000000000000 000100000000 001100000000
         66            3            5
```

The output indicates that 66 observations had no missing values, 3 had missing values for variable 4 only, and 5 had missing values for variables 4 and 5.

7.6.2 Evaluating Text as Commands

The vast majority of commands that you will present to S will either be typed in at your terminal or read from an external file through the BATCH command or the source function. However, occasions do arise when it is convenient to construct an S statement from character strings inside of S, and then to have S execute the statement as if it had been typed in or read from a file. For example, suppose you have several vectors named x.1, x.2, and so on, and you wish to create a vector of means, whose ith element is the mean for the vector x.i. It is simple to construct statements that will carry out this task as character strings in S:

```
> for(i in 1:n)
+     cat(paste("means[", i, "] <- mean(x.", i, ")",sep=""),"\n")
means[1] <- mean(x.1)
means[2] <- mean(x.2)
means[3] <- mean(x.3)
    . . .
```

To get S to use these statements as if they were input to the interpreter, the functions eval and parse can be used. First the parse function is given the text to be executed; it converts a character string into an S object of mode expression. The function eval can then be used to evaluate the expression produced by parse. The character strings displayed in the preceding example can be executed as S statements using the following code:

```
> for(i in 1:n)
+     eval(parse(text=
+         paste("means[", i, "] <- mean(x.", i, ")",sep="")))
```

The text= argument is needed because, by default, parse reads expressions from a file. Note that eval defaults to evaluating its argument in the local frame, so if you are inside a function, the statements will apply to that local evaluation frame. The local argument can be used to modify the frame in which evaluation takes place; specifically, using local=sys.parent(1) will force evaluation to take place in the frame of the calling function, whereas setting local=F will force evaluation in the global frame. In this case, any assignments that are made will take place in the default .Data directory.

7.6.3 Object-Oriented Programming

Object-oriented programming is more a new way of looking at writing programs than simply a set of new techniques. Instead of regarding your data as being passive, ready to be acted upon by a program, the object-oriented paradigm views the data as the center of attention. A program is thought of as a collection of methods, each of which knows how to deal with a particular type of data. When a new type of data is created (for example, a list containing the outcome of a type of statistical analysis), instead of modifying existing programs, new methods are created to deal with the new type of data. Although these concepts may at first seem elusive, the example that follows should help to make them more concrete. In addition, remember that if you are a user of object-oriented programs, their inner workings need not be of any interest to you. Most of the details that follow are intended for those users of S or S-PLUS who wish to write programs that allow them to take advantage of object-oriented models. For the more casual user, an understanding of the inner workings is not necessary.

To make these ideas more concrete, consider the S function `all.equal`. This function accepts two arguments, namely the two objects which are being compared, and returns a single logical value, which indicates whether or not the two objects are equal. This one function will allow you to compare scalars, vectors, matrices, character vectors, formulas, lists, results of statistical analyses, time series, and other S objects. For each of these types of objects, the definition of *equal* will be different. It's important to remember that you don't have to explicitly inform the function of the classes of the objects; one of the central ideas behind object-oriented programs is that the programs take care of that. In traditional programming, a function like this would examine the type of the object passed to the function, and use `if` statements or switches to find the appropriate code for the type of object. Let's take a look at the function `all.equal`:

```
> all.equal
function(target, current, ...)
UseMethod("all.equal")
```

The S function `UseMethod` is an indication that the program that contains it is using the object-oriented model. Such a function (like `all.equal`) is known as a generic function. In object-oriented programming, an object has an attribute known as its class. It is the class of an object that will determine which method (function) will be used when the object is passed as an argument to a generic function. The S function `class` will tell you the class of an object. In S, a method written for a particular class will consist of a function whose name is of the form `method.class`; for example, the function that determines whether two complex values are all equal is called `all.equal.complex`. Thus, when complex values are passed to the function `all.equal`, `UseMethod` will look through your search path and find an appropriate method. You can see a menu that will allow you to access the documentation for all the different methods available for a generic function by typing `?methods(functionname)`. For example:

```
> ?methods(all.equal)
The following are possible methods for all.equal
        Select any for which you want to see documentation:
1: all.equal.character
2: all.equal.complex
3: all.equal.default
4: all.equal.factor
5: all.equal.formula
```

```
 6: all.equal.language
 7: all.equal.list
 8: all.equal.lm
 9: all.equal.ms
10: all.equal.nls
11: all.equal.numeric
12: all.equal.stl
13: all.equal.versions
Selection:
```

You can then select the number of the method of interest. Once again, however, you should keep in mind that you will rarely, if ever, have to explicitly call a generic function using its full name; the object-oriented nature of the methods implemented through the **UseMethod** function takes care of figuring out what method to use and finding it in your search path.

Not every object must have a class; there will always be a default method to deal with objects that do not have an explicit class. This default method will generally be a function coded in the traditional way. You can examine the function **all.equal.default** (by typing its name) to see an example of how a default function might be coded. Some other generic functions that may be of interest include **print**, **summary**, and **plot**, as well as many of the functions introduced in Chapter 10 that are used with statistical models.

Along with the concept of classes, another key idea behind object-oriented programming is inheritability. This means that an object of a particular class may inherit the methods for some other class when a specific method does not exist. Thus, it is very easy to create a class that is very similar to an existing class, since the new class can inherit many methods from the existing one. You can inform S that an object may inherit from another class by setting its class attribute to a vector containing the name of the object's primary class, along with the names of the classes from which it can inherit. You can test whether an object can inherit from a particular class by calling the function **inherits** with the object in question as the first argument and a quoted string with the name of the class in question as the second argument.

If you create an object with a new class, all that is necessary to implement a new method for a generic function is to create a function whose name is of the form **function.class**; **UseMethod** will do the rest. If you wish to bypass the class/method system, you can remove the class of an object by passing it to the function **unclass**. Any generic functions called with the returned value of **unclass** will use the default method.

Exercises

1. The actual performance of many of the commands described in the previous sections will vary depending on the machine you are using, and the version of S (or S-PLUS) that is running on your machine. It may be of interest to run some experiments on your own computer to see the relative speeds of various implementations of commands in S.

   Use the function `unix.time` to determine how much computer time would be used by various methods of converting a covariance matrix to a correlation matrix. To convert a covariance matrix to a correlation matrix, each element of the matrix is divided by the standard deviations of its row and column. That is,

   $$corr_{ij} = \frac{covmat_{ij}}{sd_i \, sd_j} \quad \text{for } i = 1, ..., n; j = 1, ...n$$

   You can generate a random covariance matrix `covmat` of dimension `n` and a vector of standard deviations `sd` with the following statements:

   ```
   > x <- matrix(rnorm(100 * n),100,n)
   > covmat <- var(x)
   > sd <- sqrt(diag(covmat))
   ```

   Some of the methods to compare are:

   a. Direct computation using a double **for** loop
   b. Two applications of **sweep**, one dividing each row of the matrix and the other dividing each column
   c. Dividing the matrix by a conformable matrix whose *i,j*th element is $sd_i sd_j$. (*Hint:* Use the **outer** function to create the divisor matrix.)

2. Use the **apply** function, described in Section 7.1, to produce histograms of each of the variables in the **state.x77** data set. Set the value of the **mfrow** parameter appropriately so that all the plots fit on the same page. Can you think of a way to get the variable names to appear as titles on the plot? (*Hint:* Use the function **lapply** (Section 7.2) on the second element of the **dimnames(state.x77)** list.)

3. The matrix of group means for different variables calculated in Section 7.3 could also be obtained from a single call to **tapply** by using the following statements:

   ```
   > auto.data <- auto.stats[,5:8]
   > dd <- dim(auto.data)
   > auto.means <- tapply(auto.data,list(rep(cargrp,dd[2]),
   +                     rep(1:dd[2],rep(dd[1],dd[2]))),mean)
   ```

How is `tapply` processing the data to provide this result? (*Hint:* `tapply` always processes its first argument as a vector.)

4. Consider a list consisting of several matrices, each with different numbers of rows but a common number of columns. Write an S program that will create a matrix consisting of all the matrices in the list concatenated by rows.

8 Printing and Formatting

As an interactive language, S usually displays its output on a computer terminal, to be read and acted upon as statements are typed to the screen. Often it is useful to save either the input statements or the output in a file for later reference, or to create a file from data within S that will be readable by other programs. The first sections of this chapter address these issues and introduce the techniques available in S to deal with them.

A related problem is producing customized output. In addition to the usual output that functions and expressions can generate, such things as descriptive messages, tables, and other data summaries may need to be produced. The final sections of this chapter deal with these issues.

8.1 Sending Output to a File

The `sink` command can be used to reroute the output of S to a file on your computer. When you use `sink`, all of the output that would normally go to the terminal (except for prompts and error messages) will go to the file you specify as an argument to `sink`. For example, to save the output from the statements you type into a file called `my.session`, use the command

```
> sink("my.session")
```

After typing this command, you will no longer see output on the screen; it will accumulate in the file `my.session`. If you wish to have a record of the statements that you enter included in the file, you can set the S option `echo` to `TRUE` with the statement `options(echo=T)`. To restore output to the terminal, call `sink` with no arguments: `sink()`.

You should use `sink` when what you want is a transcript of your S session, with the output formatted exactly as you would normally see it on your computer screen. If you wish to write the contents of an S vector or matrix to a file for access by another program, for example, the function `write` can be used. There is one required argument to `write`: the name of the object you wish to write. By default, it will be written to a file called `data` in your current directory. If

171

the object you are writing is a matrix, keep in mind that S stores matrices in column-by-column order (Section 2.5.2). If you want a matrix to be written out row by row, pass the transpose of the matrix to **write**, using the **t** function.

Options for the write function   To override the default filename of **data**, pass a quoted string or character variable naming the file to which you wish to write, using the **file=** argument. You can control the number of data items written to each line in the file with the argument **ncolumn=**. (The default is one item per line for character variables and five items per line otherwise.) By default, **write** will overwrite the contents of a file if it already exists. If you wish to append the output of **write** to the end of the file, use the argument **append=T**. To take more control over the number of decimal places in the output written by **write**, you can use the **format** function, described in Section 8.6. Keep in mind, however, that **format** produces output of mode character, so, by default, only one item will be printed per line.

8.2   Writing S Objects for Transport

The internal representation of an S object may vary from computer to computer. The language is designed such that these differences should never concern you as a user. However, if you wish to transfer S objects from one computer to another, it is important to realize that you may not be able to simply copy the files in a **.Data** directory and transfer them to a different computer. To perform this task, S includes the functions **data.dump** and **data.restore**. These two functions write S objects in an ASCII text format that is human-readable and also computer-readable.[1] Since the ASCII characters have the same meanings on all computers, a file produced by **data.dump** will be readable by computers other than the one on which the file was made.

To use **data.dump** to write a file containing transportable versions of S objects, the function should be passed a character vector containing the names of the objects that you are dumping to the file. An optional second argument, that defaults to **"dumpdata"**, provides the name of the file that will receive the output. Often the output of the S function **objects** can serve as the first argument to **data.dump**. For example, to write all the objects in your **.Data** directory to a file called **S.trans**, you could use the following statement:

```
> data.dump(objects(),file="S.trans")
```

To restore the dumped data, pass the name of the file containing the data to the function **data.restore**. An optional second argument, **print**, if set to

---

[1] The format used by these functions has been chosen to allow rapid reading and writing of the S objects involved. If you wish to simply put an S object (like a function) into a readable format, you can use the function **dump** or **dput**. In S-PLUS, the commands **lpr** and **objprint** can be used to print objects under UNIX and Windows, respectively.

TRUE, will cause a short summary of the restored objects to be displayed as they are being restored. The objects are placed in the current working database.

8.3 Command History

Each statement you submit to S, along with additional information about the operations that S has performed, is written to a file in your .Data directory called .Audit. (The audit facility is not available under MS-DOS or Windows versions of S-PLUS; see Section 8.4 for information on command re-editing.) You can access this information through the history function of S. The history function will display or execute your previous commands, as well as allow you to edit and resubmit them.

When called with no arguments, history will display the 10 most recent commands that you have submitted to S; the optional n= argument can change the number of commands displayed. The commands are displayed in a menu, allowing you to decide which command will be executed. Typing 0 exits the menu. If the optional argument editor is set to TRUE, the default editor as specified by the option editor is opened on the selected command. Alternatively, you can specify the editing function you wish to use by setting editor to the name of that function. You can restrict the commands that are displayed to those matching a particular pattern, through the optional pattern argument to history. The menu and editing capabilities described above will then be restricted to commands containing the specified pattern.

Managing the .Audit file   Because the .Audit file contains entries for each command you submit to S, it will grow without limit over time. To control this growth, the size of the .Audit file is compared to the value of the option audit.size, which defaults to a value of 500 KB. If the .Audit file is bigger than audit.size, S prints a warning message telling you to invoke the UNIX command S TRUNC_AUDIT, which will retain only the most recent audit.size bytes of the existing .Audit file. You must issue this command from outside of your S session. Since it is only a warning, S will continue to run even if you don't truncate your .Audit file, and the file will simply continue to grow. If you wish to eliminate the existing .Audit file entirely, it is not enough to simply remove the existing .Audit file, because S will keep opening a new one. Instead, you can use the UNIX command chmod -w .Data/.Audit to make the file unwritable. This will prevent S from making entries in the file, and will conserve disk space on your system. You should truncate the file or replace it with an empty file before issuing the chmod command, to keep the space used by the file to a minimum. Once you have done this, S will always print a warning about its inability to open the audit file when you begin an S session, but this warning can be safely ignored.

8.4   Command Re-editing

When using S-PLUS under Windows, you can recall and edit previous commands by using the arrow and page keys of your computer. Invoking S-PLUS under UNIX with the **-e** option also allows you to use editor commands to recall and modify previous commands. Either vi or emacs commands may be used, depending on the value of the S option **editor**. If the environmental variable **EDITOR** is properly set, some versions of S also have this capability. Under Windows, or when you're using emacs under UNIX, no special action is needed to invoke command re-editing; when using vi under UNIX, you should press Escape before you begin command re-editing. When you type an editing command, the command to be edited will appear in the same place in which you usually type input to S; when you've completed editing the command, press Return, and the edited command will be executed. Table 8.4 shows some of the keystrokes that are useful for re-editing your commands.

Table 8.1   Keystrokes for Command Re-editing[2]

Action	Windows	UNIX using vi	UNIX Using emacs
Recall previous command	Up arrow	k	Control-p
Next command	Down arrow	j	Control-n
End of line	End	$	Control-e
Beginning of line	Home	0	Control-a
Back one character	Left arrow	h	Control-b
Forward one character	Right arrow	l	Control-f
Back one word	Control-left arrow	b	Esc b
Forward one word	Control-right arrow	w	Esc f
Delete at cursor	Delete	x	Control-d
Insert text	Type characters	i, then type characters	Type characters

8.5   Customized Printing

The basic tool for producing customized printing is the **cat** function. (See Section 3.12 for more information about **cat** and related functions and the sections below for information about printing tables.) The **cat** function accepts a variable number of unnamed arguments, along with several optional arguments described below. It converts each of the unnamed arguments to a character value and then prints that value, using the values of the named arguments to control the appearance and destination of the output.

---

[2]Control-x means press Control and the x key simultaneously; Esc x means press Escape, then press the x key.

One important use of `cat` is for printing informative messages in functions. For example, if we wished to print out a short summary regarding the size of a matrix named `x`, we could use a statement such as the following:

```
> x <- matrix(rnorm(250),50,5)
> cat("The matrix had",dim(x)[1],"rows and",
+                      dim(x)[2],"columns.\n")
The matrix had 50 rows and 5 columns.
```

Note that spaces need not be included before and after the characters in each string; by default, `cat` inserts spaces between all its arguments. However, a newline (`\n`) character must be explicitly included within a character string in order to have the output advance to a new line.

Options for the cat function   By default, the output from `cat` is displayed to the screen. If you wish the output to go to a file instead, use the `file=` argument to pass a character string with the name of the file to which you want the output to go. When using the `file=` option, you can also use the argument `append=T` to cause `cat` to put its output at the end of an existing file (instead of destroying its contents, which is the default behavior of `cat`). To automatically insert one or more characters between the unnamed arguments to `cat` in the output, use the `sep=` argument. This argument defaults to a single blank space. To eliminate any spaces between the output arguments, use `sep=""`. If `sep` is a vector, its values will be used cyclically. To have `cat` automatically insert newlines in the output, use `fill=`. If `fill` is set to `TRUE`, newlines will be inserted so that lines are no longer than the system option `width`; if `fill` is set to a numeric value, that value is used as the line width instead. Finally, the `labels` argument allows you to insert a character string of your choice at the beginning of each output line; if `labels` is a vector, its values are used cyclically.

8.6 Formatting Numbers

The display of noninteger numbers poses some special problems because of the inherent limitations of the computer's ability to store exact values for numbers that are not integers. This limitation expresses itself when you pass the `cat` function a number with potentially many decimal places, for example 25/3:

```
> cat("The answer is",25/3,"\n")
The answer is 8.33333333333333
```

In most applications, the extra decimal places are not desired. There are two simple ways to deal with this problem. If you wish to explicitly control the number of decimal places that appear, you can use the `round` or the `signif` function to convert the number to be printed into one that will appear with fewer decimal places. (See Section 3.6.4 for details.) To print the 25/3 with three decimal places, we could use this statement:

```
> cat("The answer is",round(25/3,3),"\n")
The answer is 8.333
```

An alternative is to use the function **format**, which converts its argument into a character string equivalent to that which would be produced by the default printing function of S. Thus, the number of significant digits to be printed is controlled by the system option **digits** when the **format** function is used:

```
> cat("The answer is",format(25/3),"\n")
The answer is 8.333333
> options(digits=2)
> cat("The answer is",format(25/3),"\n")
The answer is 8.3
```

If the **format** function is passed a vector of numerical values, it will produce a vector of character values whose decimal points are aligned, which often simplifies the printing of customized output. One limitation of **cat** is that it prints its arguments in the order in which it encounters them, without regard to their organization. Thus, if two vectors are passed to **cat**, the first will be displayed in its entirety before the second is displayed. But by using the **paste** function in conjunction with **cat**, neatly aligned displays of data can be produced. For example, we could display the means of the different columns of the **saving.x** data set with the following statements:

```
> snames <- dimnames(saving.x)[[2]]
> mns <- apply(saving.x,2,mean)
> mns.d <- paste("Mean for variable",snames,"=",mns)
> cat(mns.d,fill=max(nchar(mns.d)))

Mean for variable % Pop.<15 = 35.0898
Mean for variable % Pop. >75 = 2.293
Mean for variable Disp. Inc. = 1106.7862
Mean for variable Growth = 3.7576
Mean for variable Savings = 9.671
```

The **fill=** argument to **cat** is used to insure that each element of **mns.d** appears on a separate line. To produce a similar result, but with values aligned in a more attractive manner, we could use the **format** function to create vectors of character variables for the display:

```
> fsnames <- format(snames)
> fmns <- format(mns)
> mns.d <- paste("Mean for variable",fsnames,"=",fmns)
> cat(mns.d,fill=max(nchar(mns.d)))

Mean for variable % Pop.<15  =    35.0898
Mean for variable % Pop. >75 =     2.2930
Mean for variable Disp. Inc. = 1106.7862
Mean for variable Growth     =     3.7576
Mean for variable Savings    =     9.6710
```

As the above example demonstrates, `format` can also be used to pad a series of character values (`fsnames` in the above example) with blanks to make their lengths equal.

8.7 Printing Tables

In addition to graphical methods, tables are an important means of displaying information about data or summarizing a large body of information into an easily readable form. The default printing of S is often sufficient to produce attractive and easily readable tables. For example, suppose we wish to display price and repair records for all the Buicks in the data set `auto.stats`. If we simply select the information we desire, the default printing of S provides a suitable table:

```
> auto.stats[grep("Buick",dimnames(auto.stats)[[1]]),c(1,3:4)]
                 Price Repair (1978) Repair (1977)
Buick Century    4816             3             3
Buick Electra    7827             4             4
Buick Le Sabre   5788             3             4
Buick Opel       4453            NA            NA
Buick Regal      5189             3             3
Buick Riviera   10372             3             4
Buick Skylark    4082             3             3
```

One advantage of using this type of printing is that S will automatically print column dimnames for each page of the table, as defined by the system option `length`. (To suppress the repeated printing of column names, set `length` to a very large value.) In addition, if the table has more columns than can fit in a single panel (as determined by the system option `width`), row dimnames will automatically be printed at the beginning of each panel. If the data to be displayed include both numeric and character data, a data frame may be more appropriate for printing. For example, suppose we wish to print a table consisting of the minimum and maximum values for the measurements in the `auto.stats` data sets, along with the names of the cars that had these values. First we need to extract the necessary values:

```
> auto.some <- auto.stats[, 5:11]
> mins <- apply(auto.some, 2, min)
> maxs <- apply(auto.some, 2, max)
> min.name <- character(7)
> for(i in 1:7)
+       min.name[i] <-
+       dimnames(auto.some)[[1]][auto.some[, i] == mins[i]][1]
> max.name <- character(7)
> for(i in 1:7)
+       max.name[i] <-
+       dimnames(auto.some)[[1]][auto.some[, i] == maxs[i]][1]
```

Next, the quantities are combined into a data frame, and the row and column headings set with the `dimnames` function:

```
> m.tabl <- data.frame(mins,min.name,maxs,max.name)
> dimnames(m.tabl) <- list(names(mins),
+                          c("Minimum","Car with Minimum",
+                          "Maximum","Car with Maximum"))
> m.tabl
```

	Minimum	Car with Minimum	Maximum	Car with Maximum
Headroom	1.5	Datsun 200-SX	5.0	Plym Volare
Rear Seat	18.5	Amc Spirit	37.5	Volks Dasher
Trunk	5.0	Honda Civic	23.0	Merc Marquis
Weight	1760.0	Honda Civic	4840.0	Linc. Continental
Length	142.0	Renault Le Car	233.0	Linc. Continental
Turning Circle	32.0	Datsun 210	51.0	Linc. Continental
Displacement	79.0	Renault Le Car	425.0	Cad. Deville

To adjust the width of a column, you can paste blank spaces at the beginning of the appropriate column dimnames. In this example, the table would be more readable if there were additional space before the column labeled `Maximum`. We can add the space by pasting some blanks before the appropriate dimname:

```
> dimnames(m.tabl)[[2]][3] <- paste("          ",
+                                   dimnames(m.tabl)[[2]][3])
> m.tabl
```

	Minimum	Car with Minimum	Maximum	Car with Maximum
Headroom	1.5	Datsun 200-SX	5.0	Plym Volare
Rear Seat	18.5	Amc Spirit	37.5	Volks Dasher
Trunk	5.0	Honda Civic	23.0	Merc Marquis
Weight	1760.0	Honda Civic	4840.0	Linc. Continental
Length	142.0	Renault Le Car	233.0	Linc. Continental
Turning Circle	32.0	Datsun 210	51.0	Linc. Continental
Displacement	79.0	Renault Le Car	425.0	Cad. Deville

However, the default printing has some limitations. For example, the row dimname column does not have a label, and it is not possible to accommodate column labels that are longer than one line. If it is necessary to get complete control over printing, then the `cat`, `format`, and `paste` functions can be used as in the preceding section. A different version of the table above could be produced with the following statements:

```
> cat(paste(
+ format(c(" ","Statistic"," ",names(mins))),
+ format(paste(" ",c("   ","Minimum"," ",format(mins)))),
+ format(c("Car with","Minimum"," ",min.name)),
+ format(paste("        ",c(" ","Maximum"," ",format(maxs)))),
+ format(c("Car with","Maximum"," ",max.name))),
+ fill=T)
```

Statistic	Minimum	Car with Minimum	Maximum	Car with Maximum
Headroom	1.5	Datsun 200-SX	5.0	Plym Volare
Rear Seat	18.5	Amc Spirit	37.5	Volks Dasher
Trunk	5.0	Honda Civic	23.0	Merc Marquis
Weight	1760.0	Honda Civic	4840.0	Linc. Continental
Length	142.0	Renault Le Car	233.0	Linc. Continental
Turning Circle	32.0	Datsun 210	51.0	Linc. Continental
Displacement	79.0	Renault Le Car	425.0	Cad. Deville

The **format** function is used to align the values to a constant width in the columns. Note that additional spaces are added by pasting blanks to the entire column, to retain the alignment. Keep in mind that when this technique is used, column headings will not be automatically reprinted on each page.

8.8 Accessing the Operating System

The S language was designed to avoid duplication of those functions that are readily available within the operating system under which S is running. Instead, the **unix**, **dos**, and **win3** functions are provided to give access to operating system commands and to capture their output into S objects. Many S functions that deal with character data and files are simply calls to these functions, or their equivalents.

The only required argument to the **unix**, **dos**, and **win3** functions is a quoted string or character variable containing a valid command for the operating system. By default, these functions return the output of the command in a character variable, with each line of output from the operating system command producing an element in the character variable. For example, the UNIX command **wc** reports the number of lines, words, and characters in a file; its output consists of three numbers followed by the name of the file on which it has operated. Thus, assuming that the file **datafile** contains eight lines, with five numbers per line, we could get information about the file with the following command:

```
> unix("wc datafile")
[1] "       8      40      122 datafile"
```

Note that the entire output, including leading blanks, is returned as a single character value. To make output like this more useful, we could break the output of the **wc** command into individual words, inserting a newline after each one, so that each word would be an element of the output of the **unix** command. One way to achieve this goal is with the UNIX command **awk**:

```
> unix("wc datafile | awk '{for(i=1;i<=NF;i++)print $i}'")
[1] "8"        "40"        "122"        "datafile"
```

Most often these commands are embedded inside a function. In that case, the filename would be passed as a variable, and the **paste** function could be used to construct the argument to **unix**. For example, consider the task of reading a matrix from a file. If the file is organized with all the elements of a row on a separate line, we can use the **wc** command to determine the number of rows and columns in the matrix, and to read it into an appropriate S object:

```
> readmat <- function(file){
+    size <- unix(
+      paste("wc",file,"|awk '{for(i=1;i<=NF;i++)print $i}'"))
+    size <- as.numeric(size[1:2])
+    matrix(scan(file),nrow=size[1],ncol=size[2]/size[1],byrow=T)
+ }
```

Note the use of the **as.numeric** function to convert the character string produced by the **unix** command into a number.

The task of breaking a string into separate words could also be isolated into a function. The **unix** function accepts an optional argument called **input**, that can be set to an S object that will then be written to a temporary file and supplied to the UNIX command through redirection. Thus, we could write a function to break character strings into words as follows:

```
> brkline <- function (str)
+    unix("awk '{for(i=1;i<=NF;i++)print $i}'",input=str)
> brkline("This is a test")
[1] "This" "is"   "a"    "test"
```

This technique will allow the **unix** command to accept an array of character values, since they will be written to the temporary file one element per line. In this example, all the words in the array would be returned in a single vector. The function **sapply** (Section 7.2) could be used if the output from each element of the vector were desired in a separate element of a list.

The **echo** command can be used to find the value of UNIX environmental variables. To extract the name of the directory containing the files associated with S, you could use **echo** to get the value of the environmental variable **$SHOME**:

```
> shome <- unix("echo $SHOME")
```

The **paste** function could then be used to compose the filenames of specific files in the S directory.

Under MS-DOS and Windows, you can access environmental variables by using the MS-DOS **set** and **find** commands. For example, to see the name of the **HOME** environmental variable, you could use the command

```
> dos('set | find "HOME"')
```

Single quotes are used in the argument to the **dos** command because the argument to the MS-DOS **find** command must be in double quotes.

As mentioned above, the default behavior of the **unix** and **dos** functions is to return the output of the command into a character vector. In some cases, this may not be desirable or necessary. In cases like these, the optional argument **output** can be set to **FALSE**. When **output** is set to **FALSE**, these functions return the status of the command from its execution in the operating system environment. This can be especially useful when accessing the built-in UNIX function **test**. This command tests for a variety of conditions; for example, with the **-f** flag, it tests for the existence of a file. (See the online UNIX manual pages or your favorite UNIX book for more information about the **test** command.) Although the **test** command produces no visible output, it internally returns a status of zero if the named file exists, or a nonzero value otherwise. Thus, to create a function called **isfile** to return **TRUE** if a file exists and **FALSE** otherwise, we simply need to call the **unix** function with the appropriate argument, and use the logical negation operator (**!**) to convert zero to **TRUE** and nonzero to **FALSE**. This also has the useful side effect of coercing the result to a logical value. Once again, the **paste** function is used to construct the command:

```
> isfile <- function(file)!unix(paste("test -f",file),output=F)
```

Such a function would be especially useful to check for the existence of a file in a function such as the **readmat** function defined above, to avoid getting unnecessary errors from calls to the **unix** function that refer to files that don't exist. Using S-PLUS under MS-DOS or Windows, similar functionality can be achieved using the **access** command.

Exercises

1. The data set **state.region** contains a region name for each of the states in the **state.x77** data set. Create tables displaying the values of all the variables in the **state.x77** data set separately for each region. (*Hint:* Use the function **split**.)

2. Print a table from the **state.x77** data set that, for each variable, has an asterisk after the value that is a maximum for that variable. How can you make sure that the numbers still line up after you have affixed the asterisk to the maximum value?

9 Advanced Graphics

9.1 Overview

The high-level graphics commands described in Chapter 4 will produce a variety of plots with a minimum of user input. However, there are times when you need more control over the exact way in which graphics output will be arranged on a page or within a plot, or when you need to make an addition or an annotation to a simple graph. This chapter presents information on the **par** function, which allows you to set or query the value of a wide range of parameters that control the appearance of your graphical output. Low-level plotting commands whose output is added to a currently active plot are also discussed. Finally, a variety of examples of ways to combine the various different graphics commands are presented.

9.2 Graphics Parameters

To provide more complete control over the appearance of your graphical output than would be possible through the individual high- and low-level graphics commands, a set of graphics parameters are available within S. These parameters represent a wide range of facilities, including sizes of margins, positions of axis labels, scales used on axes, and the overall layout of your plots on the page or screen. You can either query the values of these parameters or change the values through the function **par**. (Note that a device must be active before you can set or examine any graphical parameters; see Section 4.1.) To query a value, call **par** with the name of the parameter in question in quotes. For example, to see the value of the parameter **usr**, which gives the limits of the current graph's coordinate system, you could use the following commands:

```
> usr.now <- par("usr")
> usr.now
[1] 0 1 0 1
```

Multiple queries can be made in one call to **par** by passing a series of quoted strings to the function, or by passing a character vector with multiple values. In either case, a list will be returned, in which the names of the elements correspond to the names of the parameters of interest. Tables 9.1 and 9.2 list the more commonly used graphics parameters, some of which will be illustrated in the examples that follow.

To set the value of a parameter, pass the new value of the parameter to **par**, using the appropriate name. For example, to set the **usr** parameter to correspond to a range from 0 to 100 in each axis, instead of the default of 0 to 1, you could use the following expression:

```
> par(usr=c(0,100,0,100))
```

You can set multiple parameters in a single call either by passing several named arguments to **par** or by passing a list whose components have the names of the appropriate parameters. In addition, many high-level plotting commands will accept named arguments corresponding to graphics parameters. For example, the graphics parameter **mgp** changes the positions of axis titles, ticks, and labels. If you change the value of **mgp** through the **par** function, the new values will be used by all plotting routines. However, if you pass the **mgp=** argument to a high-level plotting routine such as **plot**, the changes will apply only to that call to **plot**, and will not affect any other graphics commands. Note, however, that the parameters under the heading "Margins and Layouts," in Table 9.1, can be changed only through a call to **par**, even though some high-level plotting commands may accept them without a warning message.

When you set a parameter using the **par** function, the value it returns is the old value of the parameter in question (or a list containing the old values of multiple parameters). The return value from such a function call is useful for resetting parameters to their previous values—for example, when exiting a function. The following statements set the values of several parameters, and then restore them to their original values when the function exits:

```
> old.pars <- par(mfrow=c(2,2),mai=c(2,2,5,3),pch="X")
> on.exit(par(old.pars))
```

(The function **on.exit** is discussed in Section 6.10.) The technique is not restricted to use within **on.exit**—it can be used any time you need to store and reset graphics parameters. If you need to save all of the graphics parameters, you should call **par** with the argument **gr.state**. This is an S data set that contains the names of all the modifiable graphics parameters. Thus, to save the entire state of graphics parameters, you could use the statement

```
> old.state <- par(gr.state)
```

Table 9.1 Graphics Parameters, Part 1 of 2

Parameter	Description	Value
	Text Control	
adj	String justification	0 = left; 0.5 = center; 1 = right
cex	Character expansion	numeric
col	Color (device-dependent)	integer
csi	Character interline space (inches)	numeric
exp	Scientific notation for tick labels	integer (0, 1 or 2)
font	Font selection (device-dependent)	integer
pch	Plotting character	character or integer
srt	String rotation (in degrees clockwise)	numeric
	Axes	
lab	Tick intervals and label length	c(xticks,yticks,lablen)
las	Style of axis labels	0 = parallel; 1 = horiz.; 2 = perpend.
mgp	Margin line for axis title, label, and line	c(title,label,line)
tck	Tick-mark length as fraction of plot	numeric
xaxp	x-axis tick marks	c(mintick,maxtick,nintervals)
xaxs	x-axis style	"d" = direct; "e" = extended
xaxt	x-axis type	"n" = none
yaxp	y-axis tick marks	c(mintick,maxtick,nintervals)
yaxs	y-axis style	"d" = direct "e" = extended
yaxt	y-axis type	"n" = none
	Margins and Layouts (can only be set with par)	
fig	Figure region as fraction of device	c(x1,x2,y1,y2)
fin	Width and height of figure	c(width,height)
fty	Figure shape	"m" = maximum; "s" = square
mai	Margin size in inches	c(bottom,left,top,right)
mar	Margin lines	c(bottom,left,top,right)
mex	Character expansion for mar	numeric
mfg	Multiple figure position	c(rownow,colnow,nrow,ncol)
mfrow	Multiple figure layout (by row)	c(nrow,ncol)
mfcol	Multiple figure layout (by column)	c(nrow,ncol)
oma	Outer margin lines	c(bottom,left,top,right)
omd	Multiple figures as fraction of device	c(x1,x2,y1,y2)
omi	Outer margin size in inches	c(bottom,left,top,right)
pin	Width and height of plot	c(width,height)
plt	Plot region as fraction of device	c(x1,x2,y1,y2)
pty	Plot shape	"m" = maximum; "s" = square
usr	User coordinates: min and max on axes	c(xmin,xmax,ymin,ymax)

and to restore the parameters to their old values, you could use the statement

```
> par(old.state)
```

Information about the graphics parameters can be obtained by typing
help(par). Specific parameters will be referred to in the sections that follow.

Table 9.2 Graphics Parameters, Part 2 of 2

Parameter	Description	Value
	Miscellaneous	
ask	Prompt between pages or screens of output	logical
bty	Style of box around plot	character ("n" = none)
err	Error mode	-1 = don't print; 0 = print
lty	Line type (device-dependent)	1 = solid; >1 = dots and/or dashes
lwd	Line width (device-dependent)	numeric
mkh	Size of special plotting characters	numeric
new	Treat current plot as new	logical (T allows additions to plot)
smo	Smoothness parameter	numeric
xpd	Allow access outside plot region	logical
	Informational (can't be changed)	
1em	Height of a character in coordinate units	c(x,y)
cin	Width and height of char. (inches)	c(width,height)
cra	Width and height of char. (raster units)	c(width,height)
cxy	Width and height of char. (coordinate units)	c(width,height)
din	Width and height of device surface (inches)	c(width,height)
frm	Number of graph pages for current device	numeric
rsz	Width and height of a raster (inches)	c(width,height)
uin	Coordinate distance (inches)	c(x,y)

9.3   Layout of Graphics

In order to prepare graphics output exactly to your specifications, it is important to understand the way the output is composed. The overall size of the graphics output is controlled by the graphics parameter **fin**, and is a vector of length 2, giving the width and height of the output in inches. This size includes a space for a margin around the plot itself, so the size of the actual plot is controlled by the parameter **pin**, which like **fin** is a vector of length 2, giving width and height in inches. These relationships are illustrated in Figure 9.1 for the case of a single plot.

To increase the size of the margins, you can use either the parameter **mar** or the parameter **mai**. Each is a vector of length 4, giving information about the bottom, left, top, and right margins, in that order. When you use **mar**, values represent the number of lines in the margin, where a line is the size of a standard character on the output device; when you use **mai**, the values represent the size of the margin in inches. When using **mar**, keep in mind that if you add text of a larger-than-normal size in the margin, such as titles (by using the graphics parameter **cex** with a value greater than 1), you may need to increase the size of the margin to accommodate it.

One very useful feature of S is the ability to plot multiple figures on the same page or screen. Simple arrangements, using the graphics parameters **mfrow** and

Figure 9.1   Layout for a Single Plot

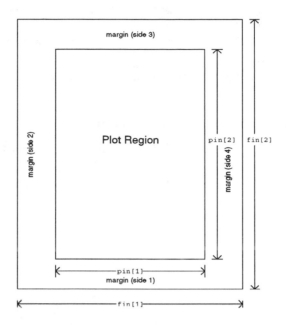

mfcol, are discussed in Section 4.3; more complex layouts are presented below and in Section 9.10.

The parameter **mfg** allows more control over the exact placement of figures on the page than is achievable using **mfrow** or **mfcol**. When you use this parameter, you can specify **nrow** and **ncol** as with **mfrow** or **mfcol**, but you can specify the position of the plot within the array explicitly, without having to produce all of the figures in the display. For example, suppose we wish to plot the two time series **rain.nyc1** and **rain.nyc2** on a full-width time series plot, and we wish to plot side-by-side displays of a normal plot and a histogram for each of the time series above and below the full-width plot. Since we want the time series plot to be wider than the other plots, we cannot use **mfrow** or **mfcol** directly, since they always assume that all plots are the same size. Instead, we can plot the first two graphs as part of a 3-by-2 array, the time series as part of a 3-by-1 array, and the final two graphs again as part of a 3-by-2 array. The statements that will achieve this are as follows:

```
> par(mfg=c(1,1,3,2))
> qqnorm(rain.nyc1,main="Normal Plot for rain.nyc1")
> par(mfg=c(1,2,3,2))
> hist(rain.nyc1,main="Histogram for rain.nyc1")
> par(mfg=c(2,1,3,1))
```

```
> tsplot(rain.nyc1,main="Plot of Rain Datasets")
> tslines(rain.nyc2,lty=2)
> par(mfg=c(3,1,3,2))
> qqnorm(rain.nyc2,main="Normal Plot for rain.nyc2")
> par(mfg=c(3,2,3,2))
> hist(rain.nyc2,main="Histogram for rain.nyc2")
```

The plots are displayed in Figure 9.2. More complex layouts can be achieved using the omd parameter, as outlined in Section 9.10.

Figure 9.2   Multiple Figures Using **mfg**

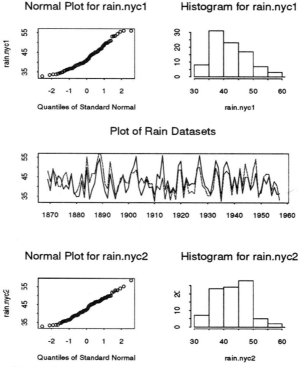

When multiple plots are to be produced on a single page (or screen), the meanings of some quantities, especially margins, change. Figure 9.3 illustrates these changes. Basically, the overall margin is now called the outer margin, and the parameters omi and oma take the place of parameters mai and mar. These parameters (mai and mar) now correspond to the inner margins associated with the current plot in the array of multiple plots. Thus, changing the value of mar will affect each of the individual plots on the page, but will not change the size of the outer margin. The parameter oma would be used to change the outer margin size.

Figure 9.3   Layout for Multiple Plots

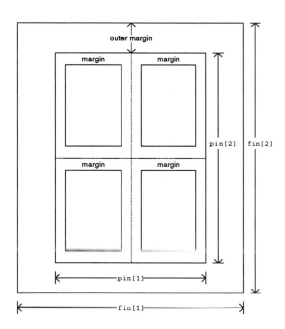

9.4   Low-Level Plotting Commands

Table 9.3 lists the low-level plotting commands available in S. These commands will not produce an entire plot on their own, but will add their output to the current plot. The use of these commands will be illustrated in the examples that follow.

Many of the low-level plotting routines require that you specify locations in terms of the x- and/or y-coordinates of the points on the current plot. (This coordinate system will be referred to as the "user coordinates" in later sections.) The user coordinates for a plot are often automatically determined by a high-level plotting routine and stored in the graphics parameter **usr**.

When your graphics device produces images directly on the screen, you can clear the device by calling the function **frame()** with no arguments. When plotting a single figure, this function removes graphics output from the screen without deactivating the device. When there are multiple figures on a page, you can use **frame()** to advance to the next figure. Repeated calls to **frame()** will eventually remove all the graphic output from the screen.

Table 9.3 Low-Level Plotting Routines

Function	Description
abline	Add regression line to plot
arrows	Draw arrow on plot
axes	Add axis labels to plot
axis	Add custom axis to plot
box	Draw box around plot
frame	Advance to next figure
labclust	Add labels to cluster plot
legend	Add legend to plot
lines	Add lines to plot
mtext	Write text in margins
perspp	Project points on perspective plots
points	Add points to plot
polygon	Draw and shade polygons
qqline	Draw median line on qqnorm plot
rug	Add data-based marks to an axis
segments	Draw disconnected line segments
stamp	Add a time stamp to a plot
symbols	Draw symbols on a plot
text	Add text to a plot
title	Add titles or axis labels
tslines	Add lines to tsplot
tspoints	Add points to tsplot

9.5   Annotating Plots with Text

The simplest tasks of adding text to plots can usually be carried out using the arguments main, xlab, and ylab, which produce an overall title, x-axis labels, and y-axis labels, respectively. Many of the high-level plotting routines will accept these arguments. When these arguments are not available, the low-level function title can be used. It accepts the following arguments: title, which is printed in large characters at the top of the plot, sub, which is printed in regular size at the bottom of the plot, and xlab and ylab, which provide labels for the x- and y-axes, respectively. You can also use the argument axes with title. If this argument is set to TRUE, tick marks, axis labels, and an enclosing box will also be added to the plot. In some cases, however, the functionality of title (or the corresponding use of main, xlab, and ylab) in a high-level plotting function will be insufficient to meet your needs. In these cases, the functions text and mtext can be used to put text anywhere on the graphics page. Use text when you wish text to appear within the plot region itself; with text, you describe the location for text using the same coordinate system as the plot uses. To place text in the margins, for example as a subtitle for a graph or for text below an axis, use mtext.

With `mtext`, the position of the text is specified with the `line=` argument, which tells how many lines away from the figure the text should be placed. To specify the margin in which the text should appear, use the `at=` argument. Values of 1, 2, 3, and 4 correspond to the bottom, left, top, and right margins, respectively. By default, lines are placed within the top margin. Noninteger and negative arguments for `line` are acceptable, providing considerable control over text in the margins.

When you have more than one plot on a page, recall that the meanings of the margins change somewhat. The description given above still holds for the individual plots that make up the page, but if you wish to put text in the overall outer margin for the entire page, you should specify `outer=TRUE` in your call to `mtext`. To accommodate text in the outer margin, you may have to adjust the outer margin size with the parameter `oma` or `omi`.

9.6 Using the Plotting Commands

Most plotting tasks require some combination of high-level and low-level commands, along with some modification to graphics parameters, either through calls to `par` or as arguments to the plotting functions themselves. The subsections that follow present a variety of graphics tasks and an explanation of how to use the commands and parameters to produce the desired display.

9.6.1 Multiple Lines or Groups of Points on a Plot

When you wish to plot several lines or points on a single plot, and the number of points in each group is the same, the function `matplot` can be used. The first two arguments to `matplot`, which are required, determine what lines or plots will be drawn. The first column of the second argument is plotted against the first column of the first argument; next, the second column of the second argument is plotted against the second column of the first argument; and so on. If there are unequal numbers of columns in the arguments, then columns from the smaller one will be recycled. In particular, as in the example below, if one of the arguments is a vector, each column of the other argument will be plotted against that vector. To illustrate, consider an experiment to test the behavior of the function `sort` when its input is an array of ordered integers, a random permutation of integers, or a random sample from the normal distribution. By using the function `unix.time`, we can determine how much computer time is used for a given task. The first element of the array returned by `unix.time` represents the time, in seconds, that the computer central processing unit (CPU) used to carry out the command, and is the quantity of interest in the experiment. The array `times` will hold the CPU times for the different sorts, using array lengths from 1000 to 10,000:

```
> times <- matrix(0,10,3)
> for(i in 1:10){
+       n <- i * 1000
+       s1 <- 1:n          # sorted
+       s2 <- sample(n)  # permutation
+       s3 <- rnorm(n)    # random normals
+       times[i,1] <- unix.time(sort(s1))[1]
+       times[i,2] <- unix.time(sort(s2))[1]
+       times[i,3] <- unix.time(sort(s3))[1]
+ }
```

To produce the plot, `matplot` is called with the vector of lengths as the first argument, and the matrix `times` as the second argument. Each column of `times` will be plotted against the length vector. To inform `matplot` as to whether lines or plots should be drawn, the optional argument `type` is used by passing a character string equal to either `"p"` for points or `"l"` for lines. If there is more than one character in the `type` argument, the individual characters will be used in order, recycling them if necessary. To specify different plotting characters, the optional argument `pch` can be used to pass a character string that will be used in the same fashion as the `type` argument. (Note that the `type` and `pch` arguments to `matplot` are single strings, not vectors of single characters.) The optional argument `lty` can be used to pass a vector of line types to `matplot`, which will be recycled as described previously. Labels and titles are added using the optional arguments `xlab`, `ylab`, and `main`. To plot the sorting times, the following command could be used; the resulting graph is shown in Figure 9.4.

```
> matplot((1:10)*1000,times,type="l",lty=1:3,
+           xlab="Length",ylab="Time (in sec.)",
+           main="Time for Sorting")
```

The functions `matlines` and `matpoints` will perform functions similar to those of `matplot`, but they add their output to the current graph instead of creating a new one.

If the sets of lines and/or points to be plotted are of different lengths, the `matplot` function cannot be used; each set of points must be plotted separately. In cases like this, it is important to make sure that the axes can accommodate all of the points in each of the groups of points to be graphed. One strategy to achieve this is to call the `plot` function with the minimum and maximum values to be plotted, but to suppress the actual plot using the argument `type="n"`. This will produce appropriate axes, but no actual plotting. Titles and axis labels are also passed to this initial call to `plot`. The actual plotting can then be done by the routine `lines` or `points`. As an example of such a plot, suppose we wish to produce a graph of fuel economy versus weight for the `auto.stats` data set, with a different smoothed line for each of the three car sizes defined by the `cargrp` variable. The commands shown below first draw the axes using the `plot` function with `type="n"`, then draw each of the three lines with a different linetype (`lty`) argument. The function `smooth.spline` returns a list containing

Figure 9.4   Multiple Lines on a Plot Using `matplot`

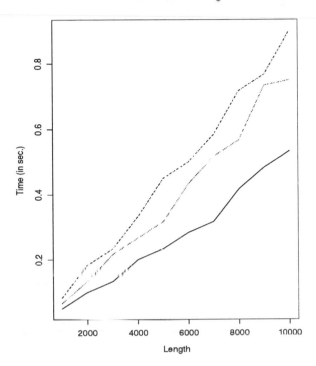

components named **x** and **y**, so its output can be passed directly to the `lines` function.

```
> cargrp <- factor(cut(auto.stats[,"Weight"],c(0,2500,3500,5000),
+               labels=c("Small","Medium","Large")))
> price <- auto.stats[,"Price"]
> mpg <- auto.stats[,"Miles per gallon"]
> plot(price,mpg,type="n",main="Fuel Use vs. Price",
+       xlab="Price",ylab="Miles per Gallon")
> j <- 0
> for(i in levels(cargrp))
+     lines(smooth.spline(price[cargrp==i],mpg[cargrp==i],spar=.01),
+           lty=j<-j+1)
```

The plot is shown in Figure 9.5.

A similar task to plotting multiple lines is plotting several groups of points on the same graph, using a different plotting character for each group. In the previous example, as an alternative to plotting a smoothed line for each group, we could produce a scatterplot with a different plotting character for each of the three sizes of cars defined by the **cargrp** variable, namely the first letter of the value of **cargrp** (S, M, or L). We could use a similar approach to the previous

Figure 9.5   Multiple Lines on a Plot

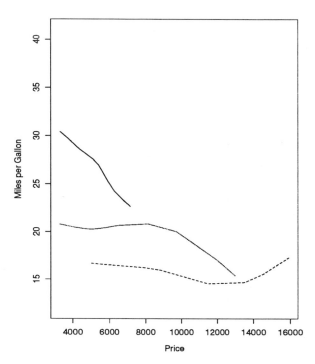

example, calling the **points** function once for each of the three groups, but an easier alternative is to use the **text** function, since it will accept an array of plotting characters:

```
> plot(price,mpg,type="n",main="Fuel Use vs. Price",
+       xlab="Price",ylab="Miles per Gallon")
> chs<-substring(cargrp,1,1)
> text(price,mpg,chs)
```

The plot is shown in Figure 9.6. Note that if the **text** command had been issued after the loop that plotted the lines in the preceding example, the points of Figure 9.6 would have been superimposed on Figure 9.5.

9.6.2 Legends

Graphs such as Figure 9.5 and Figure 9.6 can be made much more informative by including a legend that describes what each of the plotting symbols represents. To use the S function **legend**, you must first determine where you want the legend to appear, then provide the user (plotting) coordinates of either the top left corner of this location (letting **legend** determine the size of the box) or the user coordinates of the top left and lower right corners of the box. Continuing

Figure 9.6 Multiple Groups of Points on a Plot

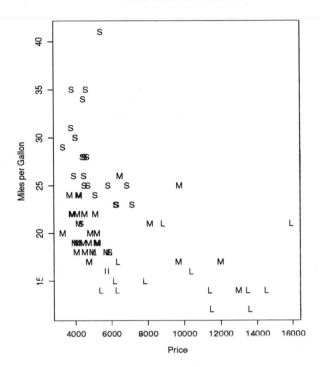

the sorting-times example, we could place a legend in the upper left-hand corner of the graph with the following statement:

```
> legend(2000,0.8,
+    c("Ordered Integers","Permuted Integers","Random Numbers"),
+    lty=1:3)
```

The parameter `lty` informs `legend` that the constructed legend should contain samples of the specified line types. The plot with the legend is shown in Figure 9.7. For the case of a plot with points, the `pch` argument to `legend` specifies which characters to display. Like the corresponding parameter in the `matplot` function, the `pch` argument to `legend` consists of a single character string containing the plotting characters to be used in the legend, not a vector of single characters. The function `paste` (see Section 3.12) can produce such a string from a vector of values if necessary. For Figure 9.6, a legend can be added with the following statement:

```
> legend(11000,40,levels(cargrp),pch="SML")
```

The graph displayed with a legend is shown in Figure 9.8.

Naturally, using the user coordinates requires examination of the specific values of a graph before the legend can be properly placed. An alternative is to reset the user coordinates using the graphics parameter `usr`, allowing you to

Figure 9.7   Multiple Line Graph with Legend

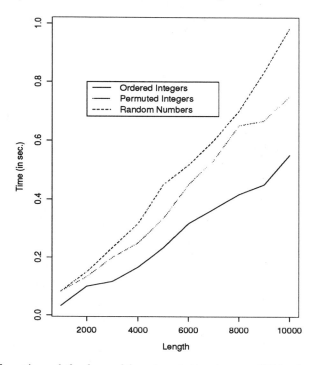

specify the location of the legend box in relative terms. (This should be the last step in creating your plot, since any other low-level graphics commands would now be using this revised set of user coordinates.) In the preceding example, similar results could have been obtained using the following statements:

```
> par(usr=c(0,1,0,1))
> legend(.6,.9,levels(cargrp),pch="SML")
```

The legend will be placed in the same position relative to the graph regardless of the magnitudes of the x and y values. In addition, the function **locator**, if it is available on your device, can provide a simple way to determine where to place a legend. See Section 4.4 for more details.

If the **legend** function cannot produce the legend you want, you can construct one yourself, using the functions **lines** and **text**.

9.6.3   Multiple Plots with Identical Axes

An alternative to putting multiple lines on a single graph is to display separate graphs, each with only one line, using the same set of axes for each of the graphs. This makes it easy to compare one graph with another. To fix the coordinates of the plots so that multiple plots have the same axes, set the graphics parameters

Figure 9.8 Multiple-Points Graph with Legend

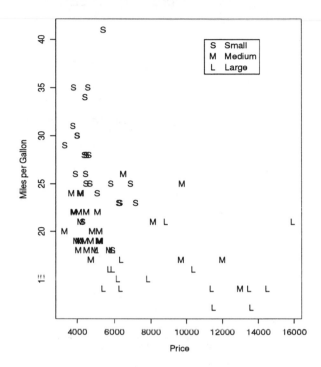

Fuel Use vs. Price

xaxs and/or **yaxs** to a value of "d". This prevents further plotting routines from recalculating the values to use on the axes. (To allow each plot to scale its own axes, set **xaxs** and/or **yaxs** to "r".) Using the sorting-times data from Section 9.6.1 as an example, we draw the axes for the first graph using the desired range, then set **yaxs="d"**. Subsequent graphs will use the same range of values for the y-axis.

```
> par(mfrow = c(3, 1))
> plot(c(1,10)*1000,range(times),type ="n",
+       xlab="Length",ylab="Time (in sec.)",
+       main = "Sorting Time for Sorted Integers")
> par(yaxs = "d")
> lines((1:10)*1000, times[,1])
> plot((1:10)*1000,times[,2],type="l",
+       xlab="Length",ylab="Time(in sec.)",
+       main = "Sorting Time for Permuted Integers")
> plot((1:10)*1000,times[,3],type="l",
+       xlab="Length",ylab="Time (in sec.)",
+       main = "Sorting Time for Random Numbers")
```

The plots are displayed in Figure 9.9.

Figure 9.9   Multiple Graphs with Fixed Axes

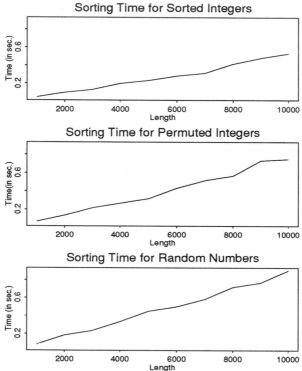

9.6.4 Special Plotting Characters

Instead of a character, the graphics parameter **pch** can be set to an integer value from 0 to 18, resulting in the special characters displayed in Figure 9.10. If you wish to access these characters for a legend, use the argument **marks=** with a vector of integers when calling **legend**. To change the size of these characters, use the graphics parameter **mkh**.

9.6.5 Logarithmic Axes

When the range of a variable is very large, it is sometimes useful to produce a graph in which the logarithm of the variable in question is plotted, instead of the variable value itself. One simple way to do this is to use the **log** function to calculate the logarithm of the variable, and to call the **plot** function in the usual manner. When you use this technique, however, the labels on the axis are in the scale of the logarithm of the variable, not in the scale of the variable itself. You can use the **log** argument to **plot** to plot the logarithm of a variable while still retaining the usual scale of the variable on the axis labels. The **log** argument to **plot** can take on the values **"x"**, **"y"**, or **"xy"**, depending on whether the x-axis, the y-axis, or both axes should be plotted logarithmically.

Figure 9.10   Special Plotting Characters Using `pch=`

0	□	7	⊠	13	⊠
1	○	8	✳	14	◺
2	△	9	⬙	15	■
3	+	10	⊕	16	●
4	×	11	⋈	17	▲
5	◇	12	⊞	18	◆
6	▽				

For example, suppose we wish to produce a plot of the logarithm of disposable income versus percentage of population over 75 in the `saving.x` data set. The commands shown below will produce two plots, one directly transforming the variable and the other using the `log` argument to plot. The graphs appear in Figure 9.11. Note that the only difference between the two plots is in the labels used along the x-axis.

```
> par(mfrow = c(2, 1))
> plot(log(saving.x[,"Disp. Inc."]),saving.x[,"% Pop. >75"],
+      main="Transformed Scale",xlab = "Log(Income)",
+      ylab="% Pop. over 75")
> plot(saving.x[,"Disp. Inc."],saving.x[, "% Pop. >75"],log = "x",
+      main="Normal Scale",xlab="Income",ylab="% Pop. over 75")
```

9.6.6 Custom Axes

The default plot command will not accept character values for either the `x` or the `y` argument, but it is often useful to produce a plot for which the axes are labeled with character strings. For example, suppose we are investigating the relationship between illiteracy rates in large and small states, and we are also

Figure 9.11   Logarithmic Plots

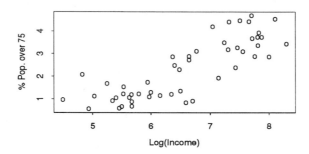

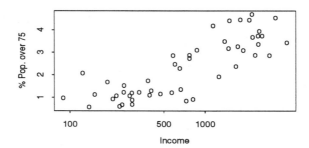

interested in the influence of the average incomes for the various groups of states. One useful way to display this information is through an interaction plot, which is a plot in which the x-axis represents one grouping variable, the y-axis is the mean or median of a statistic of interest, and separate lines are used for each level of the second grouping variable. (The S function `interaction.plot` can also be used to produce an interaction plot.) In this example, the two grouping variables are the size of the state (chosen to have three levels) and the average income of the state (chosen to have two levels). The statistic to be plotted is the mean illiteracy rate. The function `cut` is used to create categorical variables for the two grouping variables, using the quantile function to ensure that the numbers of observations in the groups are approximately equal, and `tapply` is used to calculate the means for each of the combinations of the grouping variables. The statements below produce the graph shown in Figure 9.12:

```
> area <- state.x77[,"Area"]
> income <- state.x77[,"Income"]
> area.grp <- cut(area,c(0,quantile(area,c(1/3, 2/3, 1))),
+               labels=c("Small","Medium","Large"))
> income.grp <- cut(income,c(0,quantile(income,c(1/2, 1))),
+                 labels=c("Below Median","Above Median"))
```

```
> mns <- tapply(state.x77[,"Illiteracy"],
+                list(area.grp,income.grp),mean)
> plot(c(.8,3.2),range(mns),type="n",xaxt="n",
+      xlab="Area Group",ylab = "Mean Illiteracy",
+      main="Illiteracy vs. Size for States grouped by Income")
> axis(side = 1, at = 1:3, labels = levels(area.grp))
> lines(1:3, mns[, 1])
> lines(1:3, mns[,2],lty=2)
> legend(2.25, 1.6, lty = c(1, 2), legend = levels(income.grp))
> text(2.5, 1.62, adj = 0, "Income Group")
```

In the call to **plot**, the parameter **xaxt** has been set to "n", producing a graph with no x-axis. The x-axis is then drawn using the call to **axis**; **side=1** specifies the axis on the bottom of the graph. Often when a plot like this is produced, it is visually pleasing to have some space on the x-axis before the first plotted value and after the final plotted value; thus the first argument to **plot** is c(.8,3.2), even though the values to be plotted are represented by the values from 1 to 3. Only the two extreme x values are passed to **plot**, because with type="n" no actual plotting is done, only the user coordinates are appropriately set. Similarly, the **range** function is used to provide the minimum and maximum values for the y-axis.

See the next section for information on customizing axes for barplots.

Figure 9.12   Custom Axis with Character Labels

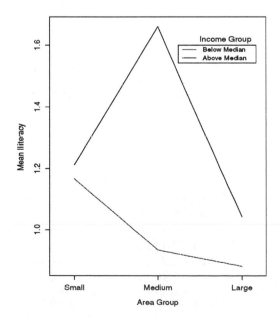

Illiteracy vs. Size for States grouped by Income

9.6.7   Customizing Barplots and Histograms

A barplot can be a useful way to display information about statistics measured for different groups. Often this information can be made more meaningful by annotating the bars, or by drawing a line above the histogram. As an example of the first case, suppose we wish to display a bar chart showing the median gas mileage for small, medium, and large cars as defined by the `cargrp` variable, introduced in Section 3.11. In addition, we would like to place a label above each bar showing the number of cars that were in the corresponding group. The `barplot` function provides a return value consisting of the x-coordinates (y-coordinates if the barplot is horizontal) of each of the bars that is drawn. Thus we can place the labels in the appropriate places using the following code:

```
> cargrp <- cut(auto.stats[,"Weight"],c(0,2500,3500,5000),
+               labels=c("Small","Medium","Large"))
> meds <- tapply(auto.stats[,"Miles per gallon"],cargrp,median)
> where <- barplot(meds,names=levels(cargrp),
+               main="Bar plot of Gas Mileage",
+               xlab="Size of Car",ylab="Mean Gas Mileage")
> text(where,meds + 1,paste("n=",table(cargrp)))
```

The plot is shown in Figure 9.13.

When a histogram is used to study the distribution of a variable, it may be useful to superimpose a line representing some theoretical distribution on top of the histogram. By using the `prob=T` argument to the `hist` function, the user coordinates of the x-axis are scaled appropriately to match the values produced by any of the density functions described in Section 3.10.1. Using the `lines` function, the output from one of the density functions can easily be used to overlay a theoretical probability distribution on top of the histogram. The technique is illustrated using the rear seat sizes from the data set `auto.stats`. Since the theoretical density has a peak that is higher than the histogram, the parameter `mar` is used to increase the space above the plot, and the `mtext` function is used to place a title on the graph. To ensure that the plotted curve covers the entire range of the histogram, the ordinates (stored in the vector `pts`) are extracted from the graphics parameter `usr`. The argument `xpd=T` allows the plotted curve to go outside the plot region, provided that it remains within the figure region.

```
> rear <- auto.stats[,"Rear Seat"]
> par(mar=par("mar") + c(0,0,3,0))
> hist(rear,prob=T,xlab="Rear Seat")
> pts <- seq(from=par("usr")[1],to=par("usr")[2],len=40)
> lines(pts,dnorm(pts,mean=mean(rear),sd=sqrt(var(rear))),xpd=T)
> mtext("Histogram of Rear Seat Size",line=4,cex=1.5)
```

The plot is displayed in Figure 9.14.

Custom axes, discussed in the previous section, are also useful for barplots, especially barplots where bars for groups are displayed side by side. To produce such plots, the argument `beside=T` is used, causing `barplot` to plot values of

Figure 9.13   Annotating a Barplot

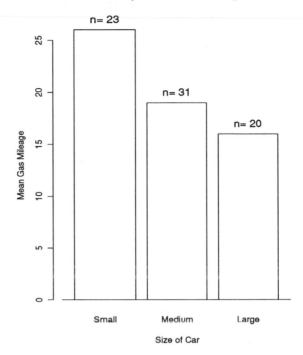

Bar plot of Gas Mileage

each row of the input matrix next to each other. The distance between the bars is controlled by the **space** argument; when **beside=T**, this argument is a vector of length 2 with the distance between bars from each row as its first element, and the distance between groups of bars as its second element. As an example, suppose we wish to produce a barplot showing the mean rating for automobiles of different sizes in the **auto.stats** data set. Because there are ratings for both 1977 and 1978, it would be informative to plot the means for the two years next to each other and separated from the bars of the other groups. The grouping variable **cargrp**, used earlier in this section, will be used to define three size groups for cars. Since the **names** argument of **barplot** expects a vector containing one value for each bar, it can be used to label the individual bars. However, to appropriately identify the groups of bars, a call to **axis** will be used.

The first step in producing the plot is to calculate the means for the two rating values for each of the size groups, using a combination of **apply** and **tapply** (see Sections 7.1 and 7.3). Since missing values will produce problems in these calculations, the **auto.stats** data set is first subsetted to remove observations with missing values:

```
> auto <- data.frame(auto.stats, Cargrp = cargrp)
> ause <- auto[!is.na(auto[, 3]) & !is.na(auto[, 4]), c(3, 4, 13)]
> rmeans <- apply(ause[, -3], 2, tapply, ause[, 3], mean)
```

Figure 9.14   Histogram with Density Curve

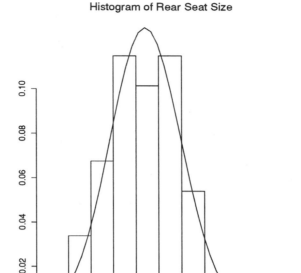

Next, the barplot is produced, using the `beside=T` and `space` arguments as described above. The x-coordinates of the bars are stored as a vector; to properly locate the labels for the groups, the means of adjacent bars must be calculated. To achieve this, the vector of x-coordinates is converted to a matrix, and the `mean` function is applied to its columns. Finally, a call to the `axis` function places the group labels on the plot. The result is displayed in Figure 9.15.

```
> xvals <- barplot(t(rmeans),beside=T,space=c(0, 1),
+                   main="Mean Repair Ratings for Car Groups",
+                   ylab="Rating",names=rep(c("1977", "1978"), 3))
> axis(side = 1,at=apply(matrix(xvals,nrow=2),2,mean),line = 1,
+       labels=dimnames(rmeans)[[1]],ticks=F)
```

9.6.8   Annotating a Perspective (3-D) Plot

The function `persp` can be used to plot three-dimensional surfaces. Data to be plotted must be in the form of a matrix, where the i, jth element represents the value of the surface to be plotted (on the z-axis) along a regularly spaced grid at the ith point along the x-axis and the jth point along the y-axis. If this matrix is given as the only argument to `persp`, values on the x- and y-axes are assumed

Figure 9.15   Side-by-Side Barplot

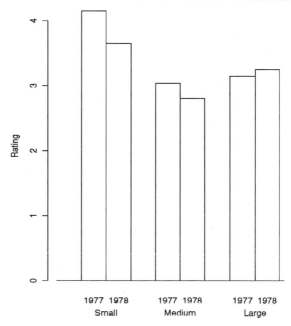

to range from 1 to the number of rows and columns in the matrix, respectively. If the first two arguments to **persp** are vectors, they are taken to be the values of x and y along the evenly spaced grid. In either case, the values to be plotted must be in a matrix as described above.

Even though the plot that is produced appears to be three-dimensional, it is actually drawn on a two-dimensional coordinate system, so annotation to the plot can be carried out in the usual way, using **lines**, **points**, **text**, and other low-level graphics functions. The information for this translation is stored in the optional output object produced by **persp**. The actual translation between the two coordinate systems is provided by the function **perspp**, which takes as its arguments the x, y, and z values in the three-dimensional system, along with the **persp** output object, and returns a list containing the x and y values in the corresponding two-dimensional system.

As an example, consider a perspective plot of the "cowboy hat" or "sombrero," as defined by a plot of

$$z = \frac{\sin(x^2 + y^2)}{x^2 + y^2}$$

for values of **x** and **y** ranging from -8 to 8. To generate the perspective plot, we first produce a matrix containing the necessary values; the function **outer** is especially useful for this task. For convenience, a function named **fff** to calculate the **z** values is used.

```
> pts <- seq(from=-8,to=8,len=50)
> fff<- function(x, y){
+        z <- sqrt(x^2 + y^2)
+        sin(z)/z }
> rmat <- outer(pts,pts,fff)
> info.p <- persp(x=pts,y=pts,rmat)
```

Suppose we wish to annotate the plot with a line pointing to the location of the maximum and a printed message. The maximum value for the function occurs near the point (0,0); however, we need to evaluate it at a point at a slight distance from (0,0) to avoid the singularity in the function. Next, a small distance is added to this point in both the x- and y-directions, and a line with an arrowhead is drawn. The function **diff** is used to get the x- and y-axis ranges from the graphical parameter **usr**. Finally, the parameter **srt** is used to ensure that the label will be parallel to the edge of the plot. The annotated plot is shown in Figure 9.16.

```
> p1<-perspp(0,0,fff(1e-10,1e-10),info.p)
> p2<-unlist(p1) + .05 *diff(par("usr"))[c(1,3)]
> arrows(p2[1], p2[2], p1$x, p1$y)
> text(p2[1], p2[2], "This is the maximum",srt=0,adj=0)
```

Figure 9.16 Annotated Perspective Plot

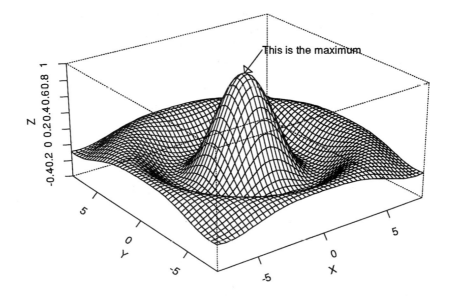

9.6.9   Drawing Diagrams

Usually a high-level graphics function will determine the user coordinates of your plot, and low-level routines can be used to add additional lines or text. However, you may occasionally want to use the graphics functions of S to produce a diagram or figure that does not have any natural x or y values from which to set the user coordinates. In cases like these, you can set the user coordinates directly through the parameter usr. By setting the xaxs and yaxs parameters to "d", you can make sure that the coordinates you set will not be changed by any of the plotting routines. To illustrate these points, consider producing a simple flow diagram, consisting of three boxes connected with arrows and an arc connecting the third box to the first box. The strategy for producing such a diagram is as follows: First set the user coordinates to some convenient value and determine the locations of the boxes and lines within that coordinate system. Then use the low-level routines symbols, arrows, and lines to produce the diagram. Since the arrows function is not designed to produce curved lines, we draw all but the last segment of the arc with lines, and finish it with a call to arrows. Note the use of the parameter cxy to ensure that the multiline text will be correctly centered within the box. The diagram is displayed in Figure 9.17.

```
> par(usr=c(0,100,0,100),xaxs="d",yaxs="d")
> symbols(list(x=rep(50,3),y=c(75,50,25)),
+          rectangles=matrix(rep(12,6),ncol=2),
+          axes=F,inches=F)
> arrows(c(50,50),c(69,44),c(50,50),c(56,31))
> text(rep(50,3),c(75,50,25) + par("cxy")[1],
+      c("Box\nOne","Box\nTwo","Box\nThree"),adj=0.5)
> z <- (0:360 * pi)/180
> x <- sin(z) * 24 + 50
> y <- cos(z) * 24 + 50
> xx <- x[x > 56]
> yy <- y[x > 56]
> lines(xx[-1],yy[-1])
> arrows(xx[2],yy[2],xx[1],yy[1])
```

9.7   Postscript Device Driver in S-PLUS

In S-PLUS, the printgraph function can be used to produce a printed copy of the graphics associated with the currently active device. Although this function can be called directly, it is generally activated through a button on a graphical display such as win.graph, motif, or openlook. Similarly, the postscript function can be called directly, but is more likely to be invoked through a printgraph call when the environmental variable S_PRINTGRAPH_METHOD is set to the value "postscript". Whether the postscript driver is called directly or through the

Figure 9.17   Simple Flow Diagram

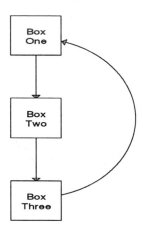

use of **printgraph**, you can alter many of the properties of your PostScript output through the S-PLUS function **ps.options**. Table 9.4 shows the arguments that can be used in this function.

The **tempfile** argument to **ps.options** is useful if there is not enough file space available in your computer's default (current) directory to accommodate the temporary file of PostScript commands that are produced by S-PLUS. The default value of **tempfile** is the character string **"ps.out.####.ps"**; the pound signs (#) are replaced with a unique number each time a plot is produced by **postscript**. To have the PostScript files written to the **/tmp** directory, for example, you could use the command

```
> ps.options(tempfile="/tmp/ps.out.####.ps)
```

The **colors** argument to **ps.options** will accept color specifications in two different forms. First, if a vector of values is given, the values define grey levels with 0 representing black and 1 representing white. To specify colors, you use a matrix with three columns; each row represents a different color, and the three columns represent values for the RGB model of color—that is the red, green, and blue intensities, respectively. Sometimes color information is presented in a different model, namely the HSB (hue, saturation, and brightness) model. If you have color specifications in this form, the function **ps.hsb2rgb** can be used to convert them to the form that **ps.options** expects.

The **paper** argument to **ps.options** accepts a wide range of different paper types as arguments. The default type is **"letter"**, which represents an 8.5-by-11 inch sheet. Other choices for this argument can be found by looking at the S object **ps.paper.regions**.

Table 9.4 Options for `postscript` Driver Through `ps.options`

Option	Description
reset	Set options to default values
tempfile	Template for naming output files
command	Command to produce printed output
horizontal	Produce printout in landscape mode
maximize	Try to maximize use of output page
paper	Size of paper
rasters	Plotting resolution (units/inch)
pointsize	Size of text in points
font	Default font number for text
setfont	PostScript procedure for setting font
fonts	Character vector of available fonts
colors	Information for color plotting
background	Color number for background
region	Numeric vector of length 4 with coordinates for lower left and upper right corners
width	Width of output in inches
height	Height of output in inches
max.vertices	Maximum number of vertices in polygons

9.8 Postscript Device Driver in S

If the `ps.options` function is not available in your version of S, you can access some of its functionality through arguments directly to the `postscript()` device driver. These arguments are summarized in Table 9.5.

The behavior of the `postscript` driver in S depends on the values of the two arguments `file` and `command`. If the argument `file` has a non-null value, then all graphics output will be directed to the file specified when the `postscript` driver is active. Thus, if you wish to use the `command` argument to specify a printing command, you should set the `file` argument to a null value. For example, to direct PostScript output to a printer named "laser" on a UNIX system, you could use the command

```
> postscript(file="",command="lpr -Plaser")
```

To send the output of the `postscript` driver to a file called `PostScript.out`, you could use the following command

```
> postscript(file="PostScript.out")
```

Note that in this case, all PostScript output will be directed to the file `PostScript.out`, until either the end of the S session, a new device driver is loaded, or the `graphics.off` function is called. If the `postscript` driver is reinvoked using the same command, then the earlier output will be rewritten.

The use of the other options is similar to that described in the previous section. In particular, the `colors` argument may be either a vector, giving a range of grey scale values, or a three-column RGB matrix.

Table 9.5 Options for `postscript` Driver in S

Option	Description
`file`	Name of file to contain PostScript output
`command`	Command to produce printed output
`horizontal`	Produce printout in landscape mode
`width`	Width of output in inches
`height`	Height of output in inches
`rasters`	Plotting resolution (units/inch)
`pointsize`	Size of text in points
`font`	Default font number for text
`preamble`	PostScript preamble
`fonts`	Character vector of available fonts
`colors`	Information for color plotting
`region`	Numeric vector of length 4 with coordinates for lower left and upper right corners
`max.vertices`	Maximum number of vertices in polygons

9.9 Multiple Graphics Devices in S-PLUS

If you invoke a single graphics device through a call to one of the graphics drivers (for example, `motif()` or `win.graph()`), then all of your graphics commands will be displayed or processed by that device. However, there are situations in which it is useful to have more than one graphics device active at one time. You may want to look at several graphs at once and view each one in a separate full-size window, or you may need to simultaneously prepare graphics output for several hardcopy devices. Table 9.6 lists some of the functions that are available within S-PLUS for managing multiple graphics devices.

Each time you invoke a graphics device, it is made the current device, to which graphics will be directed, and it is added to an internal list of such devices. You can display the list of active devices by calling `dev.list()` with no arguments. It displays the name of each active graphics device, along with its number, which may be used in other functions to specify a device. The currently active device can be displayed with `dev.cur` or set with `dev.set`. In either case, the device is referred to through its list number, as displayed by `dev.list`. The function `dev.copy` can be used to copy the current graph to either a new device, or to an existing active device whose position in the list is specified through the optional `which` argument. The function `dev.print` behaves similarly, but additionally shuts off the device to which the graph is copied. This is especially useful for printer devices such as `postscript` that must be shut off before their output can be successfully printed. A device can be manually shut off by passing its list position to the function `dev.off`. With no arguments, `dev.off` turns off the current device.

Table 9.6   Functions for Multiple Graphics Devices

Function	Use
dev.list	Display list of all active devices
dev.cur	Display currently active device
dev.off	Shut off a device
dev.set	Make a device the current one
dev.copy	Copy graphics between devices
dev.ask	Control pausing between plots
dev.next	Display next device on list
dev.prev	Display previous device on list

9.10   More Complex Layouts

The techniques described in Section 9.3 allow a great deal of flexibility in the way graphics are arranged on the graphics page or screen, but there are still times when even more control is needed. For example, suppose we wish to display time-series plots and histograms for the two New York City rain data sets, rain.nyc1 and rain.nyc2, in two rows, each consisting of a time series plot on the left and the corresponding histogram on the right. If we wanted all the plots to be of equal size, we could simply set the mfrow or mfcol parameters to c(2,2) and produce the plots. But let's assume we would like the time-series plot to take up two-thirds of the row, with the histogram using the remaining one-third. Although we can't use the mfrow, mfcol, or mfg parameter directly, we can first focus our attention on the time-series plots, producing a 2-by-1 array on two-thirds of the screen or page, then use the same technique on the remaining one-third of the page. The key to this procedure is the graphics parameter omd, which allows you to control the fractions of the device used for a number of figures. The omd parameter is set as a vector of four numbers, each between 0 and 1, representing the two fractions of the x-coordinate range to be used for the beginning and end of the figure, followed by the same two values for the y-coordinates. In the time-series example, this would mean that we could plot the time-series plots with omd set to c(0,.66,0,1) and the histograms with omd set to c(.66,1,0,1). Note that in each case we will use the entire range (0 to 1) of the y-coordinates. For the histograms, the mfg parameter is used to ensure their proper placement; the parameter new=T ensures that all the plots are placed on the same page or screen. Finally, omd is once again reset to allow titles to span both plots in each row. The resulting output is shown in Figure 9.18.

```
> par(omd = c(0, 0.66, 0, 1), mfcol = c(2, 1))
> tsplot(rain.nyc1,xlab="Year",ylab="Inches of Rain")
> tsplot(rain.nyc2,xlab="Year",ylab="Inches of Rain")
```

```
> par(omd=c(0.66,1,0,1),mfg=c(1,1,2,1),new=T)
> hist(rain.nyc1,xlab="Inches of Rain",
+      ylab="Number of Years")
> hist(rain.nyc1,xlab="Inches of Rain",
+      ylab="Number of Years")
> par(omd=c(0,1,0,1),mfg=c(1,1,2,1))
> title("Plot and Histogram for rain.nyc1")
> par(mfg=c(2,1,2,1))
> title("Plot and Histogram for rain.nyc2")
```

Figure 9.18   Complex Layout of Four Plots

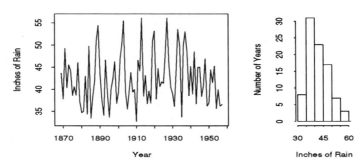

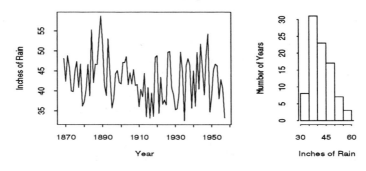

Exercises_____

1. The functions **contour** and **persp** both require a matrix representing values
   (z) to be plotted at each combination of a sequence of values of x and y.
   If you wish to produce a contour or perspective plot of data that does not
   follow such a regular pattern, it is necessary to interpolate, with the function
   interp, among the x/y combinations for which you have z values. Use the
   function **expand.grid** to create a grid of regularly spaced x and y values and
   evaluate the "sombrero" function described in Section 9.6.8 (or some other
   function) at each of these points. Then use the **sample** function to randomly
   sample points from that grid, and the **interp** function to attempt to recreate
   the values of z throughout the grid. Use **persp** and/or **contour** to display a
   plot of the interpolated values. What fraction of data is needed in the sample
   to get a good representation of the shape of the data? (*Hint*: Since the **persp**
   function will not accept missing values, you can replace missing z values from
   interp with the minimum of the nonmissing z values.)

2. Write a function that will allow placement of a legend by the use of a pointing
   device and the **locator** function. Since the **legend** function can accept either
   one or two points to define the location of the legend, make sure that your
   function allows the user to specify whether they will specify one or two points.

3. Produce a scatterplot with the origin on the right-hand side of the x-axis
   instead of its usual placement on the left-hand side. Notice that, in addition
   to negating the x values, you need to make sure that the axes display the
   appropriate labels.

10 Statistical Models

Statistical models try to explain or predict the value of one variable from the values of other, related variables. Some of the goals of modeling include accessing the relative importance of a set of variables (often called independent variables) in relation to another variable (often called the dependent variable), predicting the value of the dependent variable from those of the independent variables, or defining a mathematical relationship between the independent and dependent variables. S provides a collection of functions to find the best models for achieving these goals, along with a variety of functions to extract and display the information calculated in the process of fitting the models. The technique of object-oriented programming, described in Section 7.6.3, is used to provide a consistent interface to these functions by defining objects corresponding to each modeling technique and providing generic functions that understand the structures of these objects. Thus, regardless of the technique you use to model your data, the display and extraction of information will be the same. The modeling techniques that are available in S, and for which appropriate generic functions are provided, include:

- Linear models — `lm()` function
- Analysis of variance — `aov()` function
- Generalized linear models — `glm()` function
- Generalized additive models — `gam()` function
- Local regression models — `loess()` function
- Tree-based models — `tree()` function
- Nonlinear models — `nls()` function
- Optimization — `ms()` function

The following sections will explain the general techniques for using statistical models in S, along with specifics for some of the more commonly encountered types of analyses. In addition, functions for classical statistical inference and multivariate statistical analysis are described. It is beyond the scope of this book to cover all of the statistical background for all the methods presented, or all the features of the modeling functions themselves. The online help files, as well as statistical textbooks, should be consulted for further information.

10.1   Data for Statistical Models

Most model-building techniques make a distinction between variables that are factors (also known as categorical variables or classification variables) and those that are regressors (also known as continuous variables) in the model-fitting process. Factors are variables that represent discrete levels of a quantity, and may or may not have a logical ordering. For example, eye color, with values of blue, brown, grey, and hazel, would be a factor. In such a case, the factor is said to be unordered, since there is no natural way to rank the various eye colors. A response to a questionnaire, consisting of the choices agree, neutral, and disagree, would also be a factor, but it would be considered an ordered factor, since there is a natural ordering to the three levels: we can think of a continuum of responses ranging from agreement to disagreement, with the three levels representing points along this continuum. Because many models contain both factors and regressors, and because factors are often character variables, data frames provide a convenient means of organizing data for statistical analysis, since both character and numeric variables can be stored in a single data frame. In addition, each of the modeling functions listed above contains an argument allowing you to specify the data frame being analyzed so that the variables to be analyzed can be referred to by their names alone, without having to repeat the names of the data sets in which they are stored. Although the modeling functions will operate properly with variables stored in matrices and vectors, data frames are highly recommended for use with statistical models. If you are not familiar with data frames and anticipate using the modeling functions, you should read Section 2.5.3 to learn more about organizing your data in this format.

Once you have created a data frame, you can use either of the functions `factor` and `ordered` to set the class attributes of variables that should be treated as factors (either unordered or ordered) in future statistical analyses. For example, we could create a data frame from the `auto.stats` data matrix, adding a `Cargrp` variable corresponding to three different size groups of automobiles, with the following statements:

```
> cargrp <- cut(auto.stats[,"Weight"],c(0,2500,3500,5000))
> auto <- data.frame(auto.stats,Cargrp=cargrp)
```

To convert `Cargrp` into an ordered factor, we could use the following statement:

```
> auto$Cargrp <-
+    ordered(auto$Cargrp,labels=c("Small","Medium","Large"))
```

Any modeling function that operates on the `auto` data frame will now treat `Cargrp` as an ordered factor. Note that a similar technique can be used with variables stored as vectors, but the individual columns of a matrix have no means by which to store attributes, so the `ordered` or `factor` function would have to be called explicitly each time a given column of the matrix was used in a statistical model. This is another important reason why data frames should be your first choice when you carry out statistical modeling within S.

10.2 Expressing a Statistical Model

Statistical models in S are facilitated through the special tilde (~) operator, which can be loosely translated as "is modeled by." In the simplest case, the dependent variable appears on the left-hand side of the operator, and the independent variable on the right. To model the gas consumption (`Miles.per.gallon`) by weight (`Weight`) in the `auto` data frame described above, we could use the following formula:

    Miles.per.gallon ~ Weight

which would be read as "Miles.per.gallon is modeled by Weight." Additional terms can be added to the model formula using the plus sign (+); to add the engine displacement (`Displacement`) as an additional independent variable in the previous model, we could use the statement

    Miles.per.gallon ~ Weight + Displacement

By default, an intercept (constant) term is fitted in all models. You can fit a model without an intercept by including the term -1 in your model formula.

Single variables added to models are often referred to as main effects, especially if the variable involved is a factor. In addition to main effects, you can fit interaction terms using the colon (:) operator. Interaction terms are useful when the effect of one independent variable may change depending on the level of some other independent variable. If we felt that the effect of Displacement on Miles.per.gallon might be different depending on the size of the car (based on the `Cargrp` variable), we could add a Cargrp by Displacement interaction to the previous model by adding the term + `Cargrp:Displacement`. Alternatively, we could fit both the `Cargrp` and `Displacement` variables along with their interaction by using an asterisk (*). The term `Cargrp*Displacement` will be expanded in a model formula into the terms

    Cargrp + Displacement + Cargrp:Displacement

If more than two variables are joined by the asterisks, higher-order interactions will also be included, in addition to the main effects and two-way interactions described in the preceding example. You can limit the depth to which interactions will be included—that is, the maximum number of terms included in any interaction—by specifying the variables in question separated by plus signs and surrounded by parentheses, and followed by a caret (^) and the desired depth. For example, to fit all main effects and interactions up to depth 2 for four variables named A, B, C, and D, you could use the term (A+B+C+D)^2; it would expand to include the following terms:

    A + B + C + D + A:B + A:C + A:D + B:C + B:D + C:D

In the previous example, the values represented by `Displacement` mean exactly the same thing regardless of which level of `Cargrp` we are concerned with. It is this property of the variables that makes it reasonable to include both main effects and interactions in a statistical model. In some situations, we are interested in the different effects of one variable depending on the value of another variable, but the meaning of one of the variables changes as the level of the second variable changes. For example, suppose we were measuring the effectiveness of various teachers in several schools. It would not make sense to talk about a teacher by school interaction in addition to main effects for teachers and schools, since the teachers in one school are different from the teachers in another school. In cases like this, we say that teachers are nested within schools. To express this in a model, use the `%in%` operator, placing the variable that is nested before the operator, and the variable within which it is nested after the operator. For the teacher-and-school example, we could fit the nested effect of teachers within schools by adding the term `teacher %in% school` to a model. The slash (/) operator can be used to include both the nested effect of a variable and the main effect of the variable within which it is nested. For example, the term `school/teacher` in a model would expand to `school` and `teacher %in% school`.

Terms in models are not restricted to simple variables; any valid S expression can be used as part of a statistical model. For example, suppose we wish to model the reciprocal of gas consumption (`Miles.per.gallon`) as a function of automobile weight (`Weight`). We could use the model formula

    1 / Miles.per.gallon ~ Weight

There is no need to compute a new variable representing the reciprocal; any valid S expression can be used in a formula. When using S expressions on the right-hand side of a formula (that is, when using an expression for one of the independent variables), keep in mind that the operators listed in Table 10.1 will not have their usual meanings; they will be interpreted as special formula operators. If the terms are included as arguments to functions, they will be interpreted in the usual fashion, not as formula operators. If they are not arguments to any function, the special function `I` (identity function) can be used to protect them. For ease of reference, the table lists the usual meanings of the operators along with their meanings in model formulas.

Table 10.1 Operators Used in Model Formulas

Operator	Usual Meaning	Meaning in Formula
+	addition	Add term
−	subtraction	Remove or exclude term
*	multiplication	Main effect and interactions
/	division	Main effect and nesting
:	sequence	Interaction
^	exponentiation	Limit depth of interactions
%in%	none	Nesting

Suppose we wish to model the lengths of cars by the reciprocal of their displacement. If we do not protect the right-hand side of the formula with a call to the `I` function (or some other function), the term `1 / Displacement` will be interpreted as a nesting command; thus, the appropriate formula to use is

```
Length ~ I(1 / Displacement)
```

If we were fitting the square root of the reciprocal of the displacement, then the **sqrt** function would protect the slash from being interpreted as a special formula operator, and the following formula could be used:

```
Length ~ sqrt(1 / Displacement)
```

An alternative that eliminates the need to worry about the interpretation of special characters is to create a new variable that contains the desired transformation and use it in the formula. For example, the statement

```
> disp1 <- 1/Displacement
```

would create a new variable called **disp1**, containing the reciprocal of **Displacement**. You could then fit the previous model using the formula

```
Length ~ disp1
```

Keep in mind that the formula you use will be stored as part of the output of the modeling functions, so it is sometimes helpful to have the actual transformation in the formula for purposes of documentation.

Model formulas can be saved as S objects of the class **formula**, using the assignment operator in the usual way. For example, the following statement will save the formula modeling **Length** as the reciprocal of **Displacement** as an object called **length.formula**:

```
length.formula <- Length ~ I(1 / Displacement)
```

The object **length.formula** can then be used in any expression or function call where a formula would normally appear.

10.3   Common Arguments to the Modeling Functions

Before discussing the specifics of the individual modeling functions, it is worth noting some of the common arguments that they all accept.

10.3.1   formula Argument to the Modeling Functions

The **formula** argument is accepted by all the modeling functions mentioned in the bulleted list at the beginning of Chapter 10. For all but the function **ms** (used for optimization), the formula should have the dependent variable on the left-hand side of the tilde (~) and the independent variables on the right-hand side of the tilde, as described in the preceding section. For the **ms** function, the function to be minimized should be written as an S expression or function call on the right-hand side of the tilde, with nothing on the left-hand side.

If you specify a data frame as an argument to the modeling function (see the next subsection), you can use a period (.) on the right-hand side of the tilde to represent the additive effect of all the variables in the data frame, except for any variables referred to on the left-hand side of the formula. Since the period represents an additive set of terms, it can be followed by a caret (^) and an integer representing the depth of interactions you wish to generate. Suppose a data frame contains a dependent variable Y and three factors A, B, and C. We could fit a full interaction factorial analysis of variance using the formula

```
Y ~ A * B * C
```

or, more compactly,

```
Y ~ . ^ 3
```

Similarly, we could fit a model containing all main effects and two-way interactions with the formula

```
Y ~ . ^ 2
```

If the dependent variable does not explicitly appear on the left-hand side of the model formula, the period may be incorrectly interpreted as including the dependent variable along with the other variables in the data frame. In cases like this, you can explicitly exclude the variables in question with the minus-sign (-) formula operator. For example, suppose we are fitting an analysis-of-variance model to the data set previously described. If we create a variable called sY, which represents the square root of Y, and use it as the dependent variable in our model, the period will be incorrectly interpreted as containing Y along with the other dependent variables. This can be remedied by using the model formula

```
sY ~ . - Y
```

to explicitly tell the modeling function not to include Y as one of the independent variables.

10.3.2   data Argument to the Modeling Functions

The **data** argument to the modeling functions specifies the name of a data frame that will be used to resolve variable references in the formula that is presented to the function. This allows you to refer to variables by their names only, without the need to attach the data frame to the search list. (See Section 2.5.3 for information about attaching data frames to the search list, which is an alternative to specifying them with the **data** argument.) For example, if we wished to perform a linear regression using the **lm** function on the **Displacement** variable as a function of **Weight** in the **auto** data frame described in the previous section, we could use the following function call:

```
> auto.reg <- lm(Displacement ~ Weight, data=auto)
```

When you use a **data** argument to a modeling function you are not restricted to using only the variables found in the data frame. Any other variable that would normally be available for S expressions will be correctly resolved; however, the data frame will be searched first, before any of the members of the search list.

The **data** argument to the modeling functions will accept any S expression that evaluates to a data frame. In particular, you can specify a subsetting expression within square brackets to indicate that the model should be calculated using only part of the data frame. For example, to perform the regression of `Displacement` on `Weight` using only those observations for which the `Cargrp` variable is equal to `Small`, you could use the statement

```
> auto.reg.1 <-
+  lm(Displacement ~ Weight, data=auto[auto$Cargrp=="Small",])
```

As described in the previous section, the period (.) represents all those variables not referenced on the left-hand side of a formula when the **data** argument is used. Thus another use of the this argument is to choose a subset of variables from a data frame, so that the period will have the appropriate meaning for a particular modeling task.

10.3.3   subset Argument to the Modeling Functions

In the preceding example, it was necessary to fully qualify the `Cargrp` variable as `auto$Cargrp` because the subsetting of the data argument is carried out before S uses the data frame specified in the **data** argument for variable resolution. An alternative to the technique of subsetting through the data argument is to use the **subset** argument. You provide this argument with a logical S expression, which will be evaluated in the context of any **data** argument that is also given. Thus, identical results to the previous example could be achieved using the following statement:

```
> auto.reg.1 <-
+  lm(Displacement ~ Weight, data=auto, subset = Cargrp=="Small")
```

The **subset** argument is not available in the **nls** or **ms** function.

10.3.4   weights Argument to the Modeling Functions

In some modeling situations, it makes sense to give more consideration to some observations than to others. A single observation in a data set might actually represent many individuals, as is often the case in a survey. Some measurements may be recorded with more accuracy than others, and we may want to give added weight to those observations, using some measure such as the reciprocal of the variance as a weight. A variable or expression can be passed through the **weights** argument to apply weights in the modeling process. Keep in mind that if the technique you choose for modeling is an iterative one, specification of an expression for the weights will not cause them to be recalculated at each iteration; arguments to functions are evaluated once, when they are first referred to inside the function.

If you are using the period (.) to represent all the independent variables in a model formula, you will have to explicitly remove the variables involved in the **weights** argument if you don't want them included in the model as independent variables. See Section 10.3.1 for an example of how to specify your model in this case.

The `weights` argument is not available in the `nls` or `ms` function.

10.3.5   `na.action` Argument to the Modeling Functions

Before data are actually analyzed by a modeling function, a data frame containing all the variables to be used in the analysis is passed to the function specified through the `na.action` argument. If no such function is specified, the default action of the modeling functions is to print an error message informing you as to which variable(s) had the missing values, and to stop execution of the modeling. This default is implemented through a function called `na.fail`. A possible alternative is the function `na.omit`. This function will omit all rows (observations) of a data frame that contain missing values for any of the variables in that data frame. Another alternative is the function `na.include`, which will make `NA` a valid value for any factors in the model that have missing values. (Note that `na.include` will not solve the problem of missing values in a variable that is not a factor.)

If you have a situation that requires special handling of missing values, you can examine the functions `na.fail`, `na.omit`, and `na.include` to see how a function for handling missing values is constructed. You can also call these functions directly. For example, if you are analyzing a data set that contains missing values, and you use the `na.omit` function, remember that the data frame that is actually being analyzed has had all the rows containing any missing values for variables in the model removed. In some cases, you may need to call `na.omit` with a data frame containing the variables used in a model in order to produce a data frame compatible with the output from an analysis.

The `na.action` argument is not available in the `nls` or `ms` function.

10.3.6   `control` Argument to the Modeling Functions

For some of the modeling functions, there are a variety of parameters that can be altered to affect the modeling process. For iterative functions, such as `glm` or `nls`, these parameters may include the maximum number of iterations, or a criterion for convergence of the algorithm. In general, it will not be necessary to alter these values from the defaults, but if it is necessary, it can be done through the `control` argument of the modeling function in question.

Many of the modeling functions have a corresponding control function; for example, the control function for the `glm` modeling function is called `glm.control`. The online help file for each of these control functions will provide information as to what parameters are available. You can change the value of a parameter by specifying the new value as a named argument to the control function, and passing the function call to the modeling function through the `control` argument. For example, to increase the number of iterations in a generalized local regression model analysis from its default of 4 to 10, you could use a call such as the following:

```
> results <- loess(Y ~ A * B,
+                        control = loess.control(iterations = 10))
```

The `control` argument is not available in the `lm` or `aov` function.

10.4   Using the Statistical Modeling Objects

The statistical modeling functions should be available as soon as you start your S session; if not, they may be installed in a library (see Section 2.8). If the commands and help files for the functions are not available, type

```
> library(statistics)
```

Some of the data used in the examples below are found in the `data` library, which should be part of all S installations. To make the data available, type

```
> library(data)
```

You can then use the command

```
> library(help=data)
```

to see a brief description of the available data sets. If the functions and/or data are not available or the libraries do not exist, check with your local S administrator to make sure that the version of S or S-PLUS you are running is up to date.

Each of the modeling functions listed in the bulleted list at the beginning of Chapter 10 returns an object whose class attribute has been set to the name of the function, and which contains a variety of information concerning the modeling that has been carried out. It is generally not necessary to access the values contained in these objects directly, because a variety of generic functions are provided to allow access to the information contained in them without knowing the specifics of their contents. Nevertheless, you can find out about the components of any of these objects by requesting help for *method.object*, where *method* represents the name of the method in question, such as `lm` or `tree`. For example, the statement

```
> help(lm.object)
```

will display a description of each of the components that are present in the objects returned by the `lm` function.

More commonly, however, you can rely on the generic functions to provide you with the information you need. The two most commonly used functions in this regard are `print` (either explicitly called or used in the default printing of objects) and `summary`. Also of interest are methods provided for the function `plot`. Some of the modeling techniques have corresponding plot methods that can be used as a starting point for examination of graphical displays of a model.

In addition to the `print`, `summary`, and `plot` methods, a variety of other functions exist to extract, display, or modify information from model objects.

Some of these functions are summarized in Table 10.2. For convenience, the functions `resid` and `coef` are also provided as equivalents to `residuals` and `coefficients`, respectively.

Table 10.2 Functions for Model Objects

Function	Use
coefficients	Extract coefficients
deviance	Extract deviance
formula	Extract formula on which the model is based
fitted	Extract fitted values
pointwise	Confidence intervals on fitted values
residuals	Extract residuals
update	Modify and/or refit models

Not all model objects will provide meaningful values from all of the functions in Table 10.2. For example, no coefficients are estimated in tree-based models, so a call to the `coef` function with a tree object will return **NULL**. Specific uses of these functions will be discussed in the sections dealing with the specific modeling functions.

10.5   Linear Models and Regression (lm)

Classical linear models and linear regression are implemented in S through the `lm` function. The basic idea of a linear model is to represent the values of the dependent variable as a linear combination of terms involving the independent variables. Considering the ith observation, in which we have a dependent variable y_i and independent variables $x_{i1}, ..., x_{ip}$, we assume the following model holds:

$$y_i = \beta_0 + \beta_1 x_{i1} + \beta_2 x_{i2} + ... + \beta_p x_{ip} + e_i$$

where e_i represents the part of the observation that is unexplained by the linear combination of the data. The model is often summarized in matrix notation as

$$y = X\beta + e$$

where y represents the entire vector of dependent variable values, X is a matrix with a column for each independent variable (sometimes called the model matrix or design matrix), and e is the vector of errors (that is, the part of y that is not explained by the model).

Usually it is assumed that the unexplained component follows a normal distribution with a mean of zero and a variance (known as the residual variance) of σ^2. This variance can be estimated from the residuals, which are calculated by subtracting the fitted value based on the linear combination of βs and xs from the actual value (y_i). The lm function uses the principle of least squares; that is, it finds values of the βs that minimize the sum of the squared residuals. The residuals are also useful for a variety of diagnostic plots and measures. Because the βs always enter into the prediction equation in a linear fashion, we don't need to explicitly name them in the formula that is passed to the lm function; we need only name the independent variables that will be multiplied by the βs in the linear combination.

10.5.1 Example: Linear Regression

As an example of a regression, consider trying to estimate the fuel consumption (Miles.per.gallon) of cars in the auto.stats data set from a linear combination of Length, Weight, and Displacement. For convenience, it is assumed that the auto.stats data set has been converted to a data frame named auto, as described in Section 10.1.

```
> auto.reg <- lm(Miles.per.gallon ~ Length + Weight +
+                         Displacement, data=auto)
```

We can examine the results of the analysis by typing the name of the lm object produced by the analysis; additional information can be obtained with the summary function:

```
> auto.reg
Call:
lm(formula = Miles.per.gallon ~ Length + Weight + Displacement,
          data = auto)

Coefficients:
  (Intercept)        Length       Weight Displacement
    43.05303 -0.009540529 -0.00773257   0.01721086

Degrees of freedom: 74 total; 70 residual
Residual standard error: 3.314942
> summary(auto.reg)

Call: lm(formula = Miles.per.gallon ~ Length + Weight +
          Displacement, data = auto)
Residuals:
    Min      1Q  Median     3Q    Max
  -7.064 -1.816 -0.4751 1.402 13.65
```

```
Coefficients:
                Value Std. Error t value
(Intercept)  43.053030   6.382814  6.7451
     Length  -0.009541   0.060920 -0.1566
     Weight  -0.007733   0.002626 -2.9441
Displacement  0.017211   0.013987  1.2305
```

```
Residual standard error: 3.315 on 70 degrees of freedom
Multiple R-Squared: 0.6852
```

```
Correlation of Coefficients:
                (Intercept)  Length  Weight
     Length -0.9227
     Weight  0.5223        -0.7996
Displacement  0.0304         0.2799 -0.7713
```

The table of coefficient values in the summary output indicates that Weight is the only variable that is significantly related to Miles.per.gallon in the regression model (except for the intercept), since the absolute value of its t value is fairly large. To calculate the probability that an absolute t value this large would arise by chance, we can use the function pt to provide probabilities for each of the t values in the statement, using the residual degrees of freedom, which are 70 in this case:

```
> auto.summ <- summary(auto.reg)
> cbind(auto.summ$coefficients, Prob=
+        2 * (1 - pt(abs(auto.summ$coefficients[,"t value"]),70)))
                  Value   Std. Error    t value          Prob
(Intercept)  43.053030480 6.382813741   6.7451491 3.628635e-09
     Length  -0.009540529 0.060920454  -0.1566063 8.760059e-01
     Weight  -0.007732570 0.002626465  -2.9440975 4.392381e-03
Displacement  0.017210864 0.013987000   1.2304900 2.226326e-01
```

The multiplier of 2 before the expression involving the probability is required because the test is two-tailed; either a large positive or a large negative result will indicate that the coefficient is different from zero.

The plot method for lm produces two plots. The first is a plot of the actual value of the dependent variable versus the fitted value of the dependent variable. If the regression fitted perfectly, these points would lie on the identity line, and a dotted line representing this relationship is included in the graph. The second plot shows the absolute value of the residuals plotted against the fitted values; these points should fall in a random pattern, because one of the assumptions of the model is that the variance of the observations does not depend on the value of the dependent variable. These two plots are shown in Figure 10.1.

For the example at hand, it can be seen that one point in particular seems to be an outlier from the others, with a residual having an absolute value greater

Figure 10.1   Plots for lm Auto Example

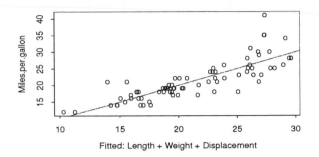

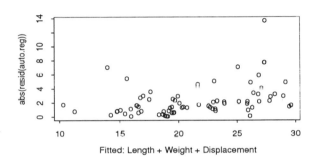

than 10. To identify this point, we can print the row name of the car in question with a statement such as the following:

```
>  row.names(auto)[abs(residuals(auto.reg)) > 10]
[1] "Volk Rabbit(d)"
```

The (d) in the name indicates that the car has a diesel engine. Since it is the only diesel engine in the data set, it might make sense to redo the regression with this point removed. One way to do so is with the **update** function:

```
> a1 <- update(auto.reg, data=auto[abs(residuals(auto.reg)) < 10,])
> summary(a1)

Call: lm(formula = Miles.per.gallon ~ Length + Weight +
        Displacement, data = auto[abs(residuals(auto.reg)) < 10, ])
Residuals:
    Min    1Q  Median    3Q    Max
 -6.759 -1.49 -0.5002 1.392 8.113
```

```
Coefficients:
                Value  Std. Error  t value
(Intercept)  38.833588    5.618559   6.9117
     Length   0.020558    0.053342   0.3854
     Weight  -0.008318    0.002287  -3.6367
Displacement  0.017876    0.012165   1.4695
```

Residual standard error: 2.883 on 69 degrees of freedom
Multiple R-Squared: 0.7203

```
Correlation of Coefficients:
             (Intercept)  Length   Weight
     Length  -0.9234
     Weight   0.5235     -0.7992
Displacement  0.0283      0.2793  -0.7707
```

In this example, the **update** function refitted the model using the same variables as the original model, but with a data set excluding the outlier. The **summary** output shows that the maximum residual has dropped from 13.65 to 8.113, and the residual standard error has dropped from 3.315 to 2.883, indicating a better fit without the outlying point. It should be emphasized that the corroborating information that the car in question had a diesel engine was essential before removal of the data could be justified.

10.6   Analysis of Variance (aov)

Analysis of variance (often abbreviated as ANOVA) follows the same general model as linear regression, and uses least squares as a criterion for determining the model parameters. The main difference between the two classes of models lies in the nature of the independent variables used. In regression, the variables are usually continuous ones, whereas in analysis of variance, the independent variables are usually factors. In addition, data for analysis-of-variance models often arise from designed experiments, wherein certain combinations of factors are studied at certain levels and not every possible combination is used.

To accommodate factors as independent variables, the techniques used for linear models cannot be used directly, because it is no longer possible to accommodate the effect of an independent variable by estimating only a single coefficient. This is because the underlying assumption that as an independent variable's value increases, its contribution to the dependent variable will either uniformly increase or uniformly decrease is not true for factors. Recall that factors may represent things such as hair color or the variety of a plant. There is no way to have a single coefficient that can explain the behavior of factors like this. Even for ordered factors, it is useful to allow the flexibility of relationships other than the strictly linear one that is implied by the use of a single coefficient in a regression. This highlights the main disadvantage of regression models: The price we pay for estimating only one coefficient per independent variable is that we have no way of expressing relationships that are not strictly linear.

Conceptually, it would be attractive to use one coefficient for each level of each factor in an analysis-of-variance model to accommodate main effects, and one coefficient for each combination of levels to accommodate interactions, but that would result in an overparameterized model matrix; that is, some of the parameters would not have unique values, and the values of the other parameters would depend on the values of the nonunique parameters. Because of this, one less coefficient than the number of levels of a factor is estimated for main effects, and the number of coefficients estimated for interactions is a product of one less than the number of levels of each factor involved in the interaction. The number of coefficients estimated to access the influence of a main effect or interaction is known as the degrees of freedom. There are a variety of ways of recoding factors to create the model matrix; see Section 11.3 for more information. For most purposes, the default method of recoding will be appropriate. Since it is understood that factors will automatically generate a number of columns in the model matrix, it is not necessary to explicitly express this in model formulas. If you wish to temporarily treat a variable as a factor, you can use the **factor** function in a model formula.

10.6.1   Example: One-Way Analysis of Variance

As an example of a one-way analysis of variance, we can test to see if the cars of several different manufacturers have significantly different mean values of fuel consumption. We will consider cars manufactured by Buick, Chevrolet, Olds, and Pontiac in the comparison. The first step is to extract the desired cars by using the **grep** function to find the desired manufacturer's name in the **auto** data frame's row names, and to create a variable called **Man** containing the manufacturer's name using the **match** function:

```
> auto.use <- auto[grep("Buick|Chev|Olds|Pont",row.names(auto)),]
> cmans <- c("Buic","Chev","Olds","Pont")
> auto.use$Man <-  factor(
+    cmans[match(substring(row.names(auto.use), 1, 4), cmans)])
```

We can now call the **aov** function to perform the analysis, saving the **aov** object in **auto.aov**, and displaying the output from both **print** and **summary**:

```
> auto.aov <- aov(Miles.per.gallon ~ Man,data=auto.use)
```

```
> auto.aov
Call:
   aov(formula = Miles.per.gallon ~ Man, data = auto.use)

Terms:
                     Man Residuals
  Sum of Squares 32.8901   238.0714
  Deg. of Freedom       3         22

Residual standard error: 3.289594
Estimated effects may be unbalanced
> summary(auto.aov)
          Df Sum of Sq  Mean Sq  F Value      Pr(F)
     Man   3   32.8901 10.96337 1.013117 0.4057513
Residuals 22  238.0714 10.82143
```

The print method for `aov` objects displays the sum of squares and degrees of freedom for each effect in the model, along with the same information for the `Residuals`, which represent the unexplained error in the model. The `Residual standard error` is the estimate of the standard deviation of this error. The message that effects may be unbalanced means that there were different numbers of observations for the different levels of `Man`; for a one-way analysis this presents no problems. The summary method displays an ANOVA table for the analysis. This is a table that has one line for each effect in the model, reporting sum of squares, mean square, F ratio, and probability of achieving an F ratio that large or larger by chance if the effect was not significant. Such a table can be produced from `lm`, `aov`, or `glm` objects; for objects other than `aov`, the function `anova` must be called explicitly to produce such a table. For this example, the table shows that the mean value of `Miles.per.gallon` is not significantly different for the different manufacturers. (To view the means themselves, the function `tapply` could be used; see Section 7.3 for details.)

The plot method for `aov` objects inherits from the class `lm`, so two plots similar to those described in Section 10.5.1 are produced by a call to `plot` with an `aov` object.

10.6.2 Example: Two-Way Analysis of Variance

We can extend the previous example to a two-way analysis by introducing a second factor based on the sizes of the cars being studied. To ensure at least one observation for each manufacturer/size combination, we create a new variable called `Cgrp`, which breaks the cars into two groups, using the median of `Weight` as the cutpoint:

```
> auto.use$Cgrp <- factor(cut(auto.use$Weight,
+              c(0,quantile(auto.use$Weight,c(.5,1))),
+              labels=c("<= Median","> Median")))
```

```
> auto.aov2 <- aov(Miles.per.gallon~Man*Cgrp,data=auto.use)
> summary(auto.aov2)
```

	Df	Sum of Sq	Mean Sq	F Value	Pr(F)
Man	3	32.8901	10.96337	1.66884	0.2092115
Cgrp	1	98.1062	98.10619	14.93371	0.0011360
Man:Cgrp	3	21.7152	7.23841	1.10183	0.3740998
Residuals	18	118.2500	6.56944		

The low probability (Pr(F)) value for Cgrp indicates that there is a difference in the mean fuel economy for the two car sizes; this, of course, is not surprising. The value for the probability corresponding to the manufacturer by size interaction (Man:Cgrp, 0.374) indicates that the interaction was not significant; such an F ratio might occur by chance as often as 37.4% of the time. This insignificant interaction means that the additive effect of car size on gas mileage was the same, regardless of manufacturer, within the variability found in the data.

10.7 Generalized Linear Models (glm)

The models encompassed by the standard linear model, both regression and analysis of variance, have certain limitations that cannot easily be overcome. First of all, the assumption of a normal distribution of errors that underlies the hypothesis tests of linear models is often not fulfilled in practice. In addition, many dependent variables cannot take on the full range of continuous values that the linear model implies. For example, dependent variables may be a fraction restricted to lie between 0 and 1, or a discrete variable such as a number of counts. Although specialized techniques have been developed to handle some of these individual cases, a large number of different techniques are unified by the concept of a generalized linear model. In matrix notation, we can extend the linear model to a generalized linear model in the following way:

$$y = g(X\beta) + e^*$$

where $g(\cdot)$ is known as the link function and e^* is an error that may come from a variety of different distributions, including the binomial, the Poisson, and the normal. By choosing an appropriate function for $g(\cdot)$ and an appropriate error distribution, a variety of different models can be accommodated. One attractive feature of generalized linear models is that the same techniques that are used to accommodate factors in the model matrix of a linear model or an analysis of variance can be used when building the model matrix for a generalized linear model, so the same types of formulas will be applicable in either case. In addition, the calculations needed to fit a generalized linear model can be carried out using a technique known as iteratively reweighted least squares, which recomputes a regular linear regression with changing weights based on the link function and the error distribution of the model.

Several different types of residuals are defined for generalized linear models, so the `residuals` function for `glm` objects has an additional argument, `type`, which can take on the values `"deviance"`, `"pearson"`, `"working"`, and `"response"`. The `"deviance"` residuals are calculated in such a way that the sum of their squared values will equal the residual deviance of the model; similarly, the sum of the squared `"pearson"` residuals is the overall χ^2 statistic. The `"working"` residual is the residual measured in terms of the linear predictor portion of the model; the `"response"` residual is simply the actual value minus the fitted value in the original scale of the variable. The default when `type` is not specified in the call to `residuals` is `"deviance"`.

10.7.1   Families, Links, and Error Distributions

The `family` argument to the `glm` function is the means by which you communicate the error distribution and link function for the model you wish to evaluate. (The `family` argument is also used in the `gam` and `loess` models, described below.) For each of the available error distributions, there is a family function that can be passed to `glm`. These functions provide a list of other functions and expressions that `glm` uses to calculate the quantities required to perform the generalized linear model analysis. The family functions corresponding to distributions that support a variety of link functions will accept an argument called `link`, specifying which link function should be used. The available family functions include `binomial`, `gaussian`, `Gamma`, `inverse.gaussian`, and `poisson`; the `quasi` family function allows you to specify a family directly through its link and variance functions and does not correspond to a particular distribution. Suitable values for the `link` argument of the family functions are `logit`, `probit`, `cloglog`, `identity`, `inverse`, `log`, `"1/mu^2"`, and `sqrt`. The online help file for `family` contains a table indicating which links can be used with which families; Table 10.3 lists the appropriate choices for the `family` argument to `glm` for a variety of common analyses.

Table 10.3 Family Argument for Common GLMs

Type of Model	family=
Regression and ANOVA	gaussian()
Logistic regression	binomial(link=logit)
Log-linear	poisson(link=log)
Constant coefficient of variation	Gamma(link=inverse)

10.7.2   Example: Logistic Regression

Logistic regression is an example of a generalized linear model with binomial errors and a logistic (`logit`) link function. The dependent variable for logistic regression can take on only two values, often corresponding to "success" and "failure." As an example, we can create a variable to measure owner's satisfaction with an automobile, called Satis, for the auto data frame created in Section 10.1, by defining "success" (TRUE) to be a rating for Repair..1978. greater than 3. We can then perform a logistic regression, using the other variables in the auto data frame. The two variables for repair records (variables 3 and 4) are removed because of their obvious correlation with the dependent variable.

```
> auto$Satis <- auto$Repair..1978. > 3
> auto.lreg <- glm(Satis ~ .,data=auto[,-c(3,4)],
+                    family=binomial(link=logit),
+                    control=glm.control(maxit=100),
+                    na.action=na.omit)
> auto.lreg
Call:
glm(formula = Satis ~ ., family = binomial(link = logit), data
        = auto[, - c(3, 4)], na.action = na.omit, control =
        glm.control(maxit = 100))


Coefficients:
 (Intercept)          Price Miles.per.gallon    Headroom
    11.68781 -2.102264e-05       0.07578372 -0.4739669


   Rear.Seat         Trunk        Weight   Length Turning.Circle
  -0.1079387 -0.09605677 -0.003977922 0.1562265      -0.807514


Displacement Gear.Ratio Cargrp.L Cargrp.Q
 0.005773379   1.914091 5.362327   1.4319


Degrees of Freedom: 69 Total; 56 Residual
Residual Deviance: 54.10299
```

The print method for glm objects gives a list of coefficient estimates and the estimate for residual deviance, which plays a similar role to the residual standard error in the classical linear model. (The summary method also includes information about residuals and correlation of estimates.) The plot method for glm is inherited from lm, and is described in Section 10.5.1.

To assess the importance of the individual factors, an analysis-of-deviance table, similar to an ANOVA table, can be displayed using the anova function. This table displays, in a sequential fashion, the deviance and residual deviance associated with each term in the glm model. If the optional argument test="Chisq" is passed to the anova function, the probability, based on the chi-squared distribution, that a change in residual deviance as extreme as the one attributed to the effect would occur by chance is also displayed. For the example at hand, we see the following results:

```
> anova(auto.lreg,test="Chisq")
Analysis of Deviance Table

Binomial model

Response: Satis

Terms added sequentially (first to last)
                 Df Deviance Resid. Df Resid. Dev    Pr(Chi)
           NULL                     68   93.89318
          Price  1  0.10152         67   93.79166 0.7500121
Miles.per.gallon  1 11.82468        66   81.96698 0.0005845
       Headroom  1  0.02836         65   81.93862 0.8662591
      Rear.Seat  1  0.33952         64   81.59910 0.5601074
          Trunk  1  0.01428         63   81.58482 0.9048859
         Weight  1  3.16206         62   78.42276 0.0753678
         Length  1  4.25695         61   74.16581 0.0390900
  Turning.Circle  1  9.00133        60   65.16447 0.0026978
   Displacement  1  0.11037         59   65.05411 0.7397277
     Gear.Ratio  1  0.57096         58   64.48314 0.4498768
         Cargrp  2 10.38015         56   54.10299 0.0055716
```

It can be seen from the table that not all of the effects are significantly related to satisfaction through the logistic model. Techniques for stepwise selection of models are outlined in Section 11.7.

When the effects involved in a logistic regression are all factors, it is not uncommon to display the data in the form of a table, showing the numbers of "successes" and "failures" in separate columns for each level of the factor (or factors) in question. For models like this, the dependent variable presented to glm should be a matrix with two columns, one representing the number of successes, and the other the number of failures. For example, we might be presented with a summary table for satisfaction as related to the size of cars (Cargrp), looking like this:

	Not Satisfied	Satisfied	Total
Small	6	16	22
Medium	21	6	27
Large	13	7	20

To fit a logistic model, once again using the satisfaction variable as a dependent variable, we could arrange the data from the table in the following way:

```
> satis <  matrix(c(6,21,13,16,6,7),ncol=2)
> satis
     [,1] [,2]
[1,]    6   16
[2,]   21    6
[3,]   13    7
> grp <- factor(c("Small","Medium","Large"))
> tbl.lreg <- glm(satis~grp,family=binomial(link=logit))
> anova(tbl.lreg,test="Chisq")
Analysis of Deviance Table

Binomial model

Response: satis

Terms added sequentially (first to last)
     Df Deviance Resid. Df Resid. Dev      Pr(Chi)
NULL                     2    13.60927
  grp  2 13.60927        0     0.00000 0.001108626
```

The analysis-of-deviance table shows that the grp variable is highly significant, indicating that a strong relationship between satisfaction and auto size exists using the logistic model. The deviance reported for the grp variable in the above example is the same as would be obtained through the following model:

```
> glm(data=auto,Satis~Cargrp,family=binomial(link=logit),
+      na.action=na.omit)
```

where Satis is a binary variable representing the owner satisfaction ratings for the individual cars.

10.7.3 Example: Log-Linear Model

As an example of a log-linear model, the market.survey data frame from the data library (see Section 10.4) will be used. Among the variables in the data set are age, which tells us to which of several age groups respondents belong; income, which tells us to which of several income groups respondents belong; and card, a binary variable that indicates whether or not the respondent has a telephone calling card. As a first example, we will consider the relationship between age and income in the sample of consumers who were questioned. Because the log-linear model is modeling the number of counts as a function of a set of independent variables, the data must first be summarized into a table of counts, using the function table. Next, the row and col functions are used as indices into the dimnames vectors from the output of table to create a data frame suitable for analysis:

```
> m.tabl <- table(market.survey$age,market.survey$income)
> m.tabl
        <7.5 7.5-15 15-25 25-35 35-45 45-75 >75
18-24    10    11    20    12     5    1    0
25-34    17    27    52    62    26    7    5
35-44    12    19    35    45    30   29    5
45-54     7    14    30    21    24   19    8
55-64    17    22    23    24    15   10    7
  65+    33    20    25     7     7   14    6
> rnames <- dimnames(m.tabl)[[1]]
> cnames <- dimnames(m.tabl)[[2]]
> m.frame <- data.frame(Age=rnames[as.vector(row(m.tabl))],
+                 Income=cnames[as.vector(col(m.tabl))],
+                 Count=as.vector(m.tabl))
> m.frame
      Age Income Count
1 18-24   <7.5    10
2 25-34   <7.5    17
3 35-44   <7.5    12
4 45-54   <7.5     7
5 55-64   <7.5    17
             . . .
```

The glm function can now be used with a log link and Poisson error term (**family = poisson(link = log)**):

```
> market.ll <- glm(Count~Age*Income,data=m.frame,
+                         family=poisson(link=log),
+                         control=glm.control(maxit=50))
> anova(market.ll,test="Chisq")
Analysis of Deviance Table

Poisson model

Response: Count

Terms added sequentially (first to last)
            Df Deviance Resid. Df Resid. Dev      Pr(Chi)
     NULL                     41    377.1757
      Age  5  95.8977         36    281.2780 0.000000e+00
   Income  6 161.5714         30    119.7066 0.000000e+00
Age:Income 30 119.7066         0      0.0000 1.142531e-12
```

The **control** argument to glm was used because a preliminary analysis indicated that convergence was not met with the default value of **maxit=10**. The analysis of deviance shows that both main effects, as well as the interaction, were significant, indicating that there were differing numbers of respondents in the different age and income groups (main effects) and that the distributions of numbers of respondents in the different income groups were different, depending on which age group was selected. (The interaction could equivalently be interpreted reversing the role of age and income.)

Oftentimes, data for log-linear models are found in the form of a table, showing the levels of the factors and the counts for each cross-classification. For example, the counts for the market.survey data using age, card, and income as factors might appear as follows:

Age	Card	<7.5	7.5–15	15–25	25–35	35–45	45–75	>75
				Income				
18–24	No	8	10	14	9	2	1	0
	Yes	1	1	4	2	3	0	0
25–34	No	15	21	33	41	13	3	5
	Yes	2	4	16	17	13	4	0
35–44	No	11	15	28	32	20	14	1
	Yes	1	4	6	11	10	15	4
45–54	No	6	11	24	15	19	12	5
	Yes	1	3	6	6	5	7	3
55–64	No	13	17	16	16	9	6	4
	Yes	4	5	7	8	0	4	3
65+	No	26	14	16	6	6	6	3
	Yes	7	6	9	1	1	8	3

One solution would be to enter the value of each of the factors, followed by the corresponding entry of the counts from the table, but this will generally be quite tedious. Since the table imposes a regular structure on the data, we can read in the counts as a vector, and generate the appropriate levels of the factors using the rep function. One way to do this is to use the function gen.levels, which appears below. This function accepts four arguments. The first argument is the number of distinct levels of the factor in question, the second is the number of repetitions desired for each level, and the third is the length of the final output vector. The fourth argument, which is optional, is a vector of values to be used in the output vector instead of the default of consecutive integers. In the table above, assume we read the data into an S object called counts in a row-by-row fashion with the following call to scan:

```
> counts <- scan()
1: 8 10 14 9 2 1 0 1 1 4 2 3 0 0 15 21 33 41 13 3 5
22: 2 4 16 17 13 4 0 11 15 28 32 20 14 1 1 4 6 11 10 15 4
43: 6 11 24 15 19 12 5 1 3 6 6 5 7 3 13 17 16 16 9 6 4
64: 4 5 7 8 6 4 3 26 14 16 6 6 3 7 6 9 1 1 8 3
85:
```

The variable Age has six levels; to correspond to the row-by-row ordering of the table, we need to repeat each level 14 times. Similarly, Card has two levels, each of which needs to be repeated seven times. Finally, Income has seven levels, each repeated once. For all variables, the length will be the same as the length of counts, which is 84. The source for the gen.levels function, and the statements to create a data frame from the tabled data, are as follows:

```
> gen.levels <- function(n1, n2, l, vals = seq(to = n1))
+{
+        if(length(vals) != n1)
+                stop("Wrong number of values.")
+        nn <- 1/(n1 * n2)
+        if(nn != 1 %/% (n1 * n2))
+                stop("l is not a multiple of n1 and n2.")
+        vals[rep(rep(1:n1, rep(n2, n1)), nn)]
+}
> mtable.frame <- data.frame(
+   Age = gen.levels(6,14,84,
+              c("18-24","25-34","35-44","45-54","55-64","65+")),
+   Card = gen.levels(2,7,84,c("No","Yes")),
+   Income = gen.levels(7,1,84,
+     c("<7.5","7.5-15","15-25","25-35","35-45","45-75",">75")),
+   Counts = counts)
```

The mtable.frame data frame is suitable for analysis using glm with family = poisson(link = log):

```
> mtable.ll <- glm(Counts ~ Age*Card*Income,data=mtable.frame,
+                  family=poisson(link=log),
+                  control=glm.control(maxit=50))
> anova(mtable.ll,test="Chisq")
Analysis of Deviance Table

Poisson model

Response: Counts

Terms added sequentially (first to last)
                 Df Deviance Resid. Df Resid. Dev    Pr(Chi)
        NULL                       83    559.5742
         Age  5   93.5863          78    465.9879 0.0000000
        Card  1  142.1609          77    323.8270 0.0000000
      Income  6  150.0609          71    173.7662 0.0000000
    Age:Card  5    3.9013          66    169.8649 0.5637172
  Age:Income 30  114.1038          36     55.7612 0.0000000
 Card:Income  6   30.8019          30     24.9593 0.0000277
Age:Card:Income 30 24.9593          0      0.0000 0.7270075
```

All main effects are significant, as are two of the two-way interactions.

10.8 Generalized Additive Models (gam)

Although the use of a link function and a variety of error distributions greatly enhances the range of data that can be accommodated by a linear model, the linearity at the core of the model formulation creates severe restrictions in the range of independent/dependent variable relationships that can be effectively studied. In some cases it is possible, on theoretical grounds, to use a transformation of an independent or dependent variable to partially alleviate the problem, but in other cases, nonlinearities in these relationships may be difficult or impossible to state analytically. The class of models known as generalized additive models overcomes this difficulty by modeling a dependent variable in the following way:

$$y_i = g(\alpha + f_1(x_1) + f_2(x_2) + ... + f_p(x_p)) + e_i^*$$

where the $f_i(\cdot)$s are functions determined by the nature of the relationship of the x_is to the y_is, $g(\cdot)$ is a link function as in a generalized linear model, and the e_i^*s represent the error in the model. Thus, there is no need to know the exact analytical form of the transformation of the independent variables used in the additive portion of the model; the modeling process finds those transformations that are most appropriate and the nature of the relationships can be viewed using graphical techniques. When a model formula is expressed for a generalized additive model, variables that are to enter linearly (that is, variables for which the $f_i(x_i)$ is just $\beta_i x_i$) are entered in the formula in the usual fashion. For those variables for which the function $f_i(\cdot)$ needs to be estimated, one of a variety of smoothing functions should be called. These include s, for producing spline smoothes; bs, for B-spline smoothing; ns, for natural cubic spline smoothing; and lo, for a lowess (locally weighted regression) smooth. These functions do not actually perform the smoothing; they only prepare the data for the algorithm used in gam model fitting. (The functions loess.smooth and smooth.spline are available for smoothing data outside the context of a gam model.)

10.8.1 Example: General Additive Model

The use of the gam function will be illustrated with the stack loss data found in the S matrices stack.x and stack.loss. (Type help(stack) for more information about these data.) First the variables are combined in a data frame and are assigned meaningful names:

```
> stack<-data.frame(cbind(stack.x,stack.loss))
> names(stack)<-c("AirFlow","WaterTemp","AcidConc","Loss")
```

The dependent variable is Loss, indicating the percentage of ammonia (times 10) lost from a chemical plant. The three independent variables are measurements related to the process generating ammonia in the plant. The model we will fit is a Gaussian model using spline smoothing on the independent variables. Since the

Gaussian error distribution with identity link is the default for **gam**, no **family** argument need be supplied.

```
> stack.gam <- gam(Loss~s(AirFlow)+s(WaterTemp)+s(AcidConc),
+                       control=gam.control(bf.maxit=50),data=stack)
> stack.gam
Call:
gam(formula = Loss ~ s(AirFlow) + s(WaterTemp) + s(AcidConc),
        data = stack, control = gam.control(bf.maxit = 50))

Degrees of Freedom: 21 total; 8.00097 Residual
Residual Deviance: 67.79171
> summary(stack.gam)

Call: gam(formula = Loss ~ s(AirFlow) + s(WaterTemp) + s(AcidConc),
        data = stack, control = gam.control(bf.maxit = 50))
Deviance Residuals:
       Min        1Q     Median        3Q       Max
 -3.089759 -1.604992 0.2439517 0.8764971 3.967667

(Dispersion Parameter for Gaussian family taken to be 8.472936 )

    Null Deviance: 2069.238 on 20 degrees of freedom

Residual Deviance: 67.79171 on 8.00097 degrees of freedom

Number of Local Scoring Iterations: 1

DF for Terms and F-values for Nonparametric Effects

               Df Npar Df   Npar F      Pr(F)
  (Intercept)   1
    s(AirFlow)  1         3 0.934405 0.4676402
 s(WaterTemp)   1         3 3.171167 0.0851828
  s(AcidConc)   1         3 0.975555 0.4507614
```

The control argument in the **gam** call sets the parameter **bf.maxit** to 50 in response to preliminary runs of the model. This parameter determines how many interactions to allow in the backfitting phase of the modeling process—that is, the phase in which the smoother is fitted to the data. In addition, there is a second iteration parameter, **maxit**, which controls the number of iterations in the overall fitting phase. The print method for **gam** objects reports the call, the degrees of freedom of the model, and the residual deviance, which for the Gaussian model is equivalent to the residual sum of squares. The summary method reports some quantiles of the deviance residuals and an analysis-of-deviance table presenting the result of the fit. The degrees of freedom reported are actually an approximation, since no parameters are actually being estimated as they would be in a regression or an ANOVA model. The plot method for **gam** objects shows the smooth curve applied to each of the independent variables; if the optional argument `ask=T` is passed to `plot` with a **gam** object, a menu displaying possible graphics choices is displayed. The statements

```
> par(mfrow=c(3,1))
> plot(stack.gam)
```

produce the three plots shown in Figure 10.2.

The plots indicate that there does not seem to be a single consistent linear relationship between the independent variables and the dependent variables throughout the entire range of independent-variable values. The tick marks above the x-axis (known as a rug) indicate the locations of the actual data points.

Figure 10.2   Plots for **gam** Stack Loss Example

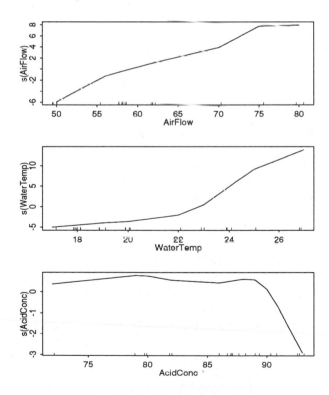

10.9 Local Regression Models (loess)

In the usual regression setting, one set of coefficients is used to model the relationship between an independent variable and a dependent variable throughout the entire range of the independent variable's values. The concept of local regression extends this usual regression model by recalculating the best-fitting regression model in several of a series of intervals that span the range of the dependent variable(s). These intervals are known as neighborhoods. These models assume that, local to a neighborhood of independent variable values, the relationship between the dependent and independent variables can be well explained by either a linear or a quadratic regression. Thus, in their simplest form, loess models are very similar to gam models using local regression smoothers. However, gam models are strictly additive, whereas loess models, which use quadratic regression, can accommodate more complex regression functions by including terms that are functions of the products of pairs of independent variables. In addition, loess models can be specified with a family argument used in the same way as a glm or gam model, further increasing their flexibility.

10.9.1 Example: Local Regression Model

The stack data frame created in Section 10.8.1 will be used to illustrate a local regression model. The two variables AirFlow and WaterTemp will be used to model the value of Loss. The statement to fit the model is as follows:

```
> stack.loess <- loess(Loss ~ AirFlow * WaterTemp,data=stack)
> stack.loess
Call:
loess(formula = Loss ~ AirFlow * WaterTemp, data = stack)

    Number of Observations:           21
    Equivalent Number of Parameters: 8.3
    Residual Standard Error:          3.411
    Multiple R-squared:               0.95
    Residuals:
        min  1st Q median 3rd Q    max
     -2.896 -1.133 0.4964 1.264 6.104
```

The print method for loess objects provides information about the equivalent number of parameters, residual standard error, and multiple R-squared, and a display of the quartiles of the residuals from the fit. The summary method displays the same information, but can be assigned to a value to make the quantities available for further calculation.

The plot method for loess objects varies depending on the number of independent variables in the model. When there is only one independent variable, a smoothed line plot of the fitted values versus the independent variable is displayed. For two independent variables, two coplots are produced (see Section 11.1 for a description of a coplot). Each of the coplots displays fitted values versus one of the independent variables in the panels of the coplot, and uses the remaining independent variable as the given variable, fixing it at several values throughout its range. For three dependent variables, similar plots are produced using the two independent variables as given variables. For more than three independent variables, there is no plot method for `loess` objects.

The two coplots produced by the statement

```
> plot(stack.loess)
```

are shown in Figure 10.3. From the plots, it can be seen that the effect of either of the independent variables changes dramatically over the range of the other independent variable. In other words, there is a strong interaction between `AirFlow` and `WaterTemp`. In particular, from the first coplot it can be seen that for low values of `WaterTemp`, `AirFlow` has a much more pronounced effect on `Loss` than it does when `WaterTemp` is at a higher value. This same effect can be seen from a different point of view in the second coplot. This example and the previous one illustrate the shortcomings of trying to use classical linear regression techniques on data such as these.

Figure 10.3 Plots for **loess** Stack Loss Example

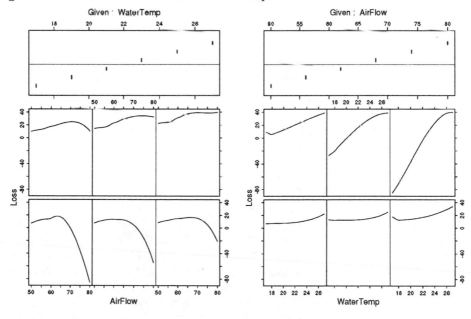

10.10   Tree-Based Models (tree)

As an alternative to the techniques described above, tree-based models can be used either for prediction (similar to a regression analysis) or for classification. They use a principle known as binary recursive partitioning to achieve this goal. Basically, at each step of the tree-building process, the values of the dependent variables are examined for all possible splits of the data to find the split that most effectively separates the dependent variable into homogeneous groups. For continuous independent variables the splits are defined by a single value: an observation goes into one node if its value is less than or equal to the split value, and into another node if its value is greater than the split value. For factors, all possible partitions of the levels into two nonoverlapping groups are considered, and observations are split based on which group contains their value for the factor on which the split is based. Because of the lack of assumptions, these models perform well in cases where more parametric models might not be effective. Because of the nature of the partitioning process, the only terms allowed in **tree** models are additive ones; each variable entered in the model may potentially be used to define the splits at each stage of the modeling process.

10.10.1   Example: Tree-Based Model for Classification

We will once again consider the **Satis** variable, created from the **auto** data frame in Section 10.7.2, to construct a tree-based model. As before, the repair ratings are eliminated from consideration from the model. For the **tree** function to perform classification instead of regression, the dependent variable must be a factor, so **Satis** is temporarily converted to a factor for the analysis. We can perform the analysis with the following statements:

```
> auto.tree <- tree(factor(Satis) ~ .,data=auto[,-c(3,4)],
+                     na.action=na.omit)
> auto.tree
node), split, n, deviance, yval, (yprob)
        * denotes terminal node

 1) root 69 93.890 FALSE ( 0.5797 0.4203 )
    2) Turning.Circle<39.5 28 29.100 TRUE ( 0.2143 0.7857 )
      4) Weight<2365 20 24.430 TRUE ( 0.3000 0.7000 )
        8) Miles.per.gallon<29.5 13 17.940 TRUE ( 0.4615 0.5385 )
          16) Price<4672 7   8.376 FALSE ( 0.7143 0.2857 ) *
          17) Price>4672 6   5.407 TRUE ( 0.1667 0.8333 ) *
        9) Miles.per.gallon>29.5 7   0.000 TRUE ( 0.0000 1.0000 ) *
      5) Weight>2365 8   0.000 TRUE ( 0.0000 1.0000 ) *
    3) Turning.Circle>39.5 41 37.480 FALSE ( 0.8293 0.1707 )
      6) Length<209.5 27   0.000 FALSE ( 1.0000 0.0000 ) *
            . . .
```

```
> summary(auto.tree)
```

```
Classification tree:
tree(formula = factor(Satis) ~ ., data = auto[, - c(3, 4)],
        na.action = na.omit)
Variables actually used in tree construction:
[1] "Turning.Circle"   "Weight"              "Miles.per.gallon"
[4] "Price"            "Length"
Number of terminal nodes:  7
Residual mean deviance:  0.4546 = 28.19 / 62
Misclassification error rate: 0.08696 = 6 / 69
```

The print method for **tree** objects describes each split that was performed on the data, reporting the number of observations in the resulting node of the split and the deviance of the node. For tree-based models, the deviance is a measure of homogeneity within the node; small values indicate very similar values. For a classification model, such as this example, the most commonly occurring value of the dependent variable is displayed, along with a vector showing the distribution of observations in each node with regard to the dependent variable. So, for example, node 9, which was formed by splitting on the condition **Miles.per.gallon>29.5**, contained seven observations; the deviance was zero, because all of those observations had a value of **TRUE** for the **Satis** variable. An asterisk at the end of a line means that this represents a terminal node—that is, one that was actually used in the final decision-making process.

The summary method displays the call that produced the tree, a list of variables actually used, and some brief summaries of the effectiveness of the tree.

The plot method for **tree** objects is probably the most useful tool for viewing the results of a tree-based analysis. It displays a tree diagram, showing in a hierarchical fashion each split that took place. In addition to the plot method, there is a text method for **tree** objects, which prints information about the split and the dependent variable's values in each node. For the example, the diagram, which is displayed in Figure 10.4, can be produced by the following commands:

```
> plot(auto.tree)
> text(auto.tree)
```

From the diagram, we can see that the first split was based on **Turning.Circle<39.5**; this produced two nodes. The left node was further split based on the value of **Weight**, the right node was split based on **Length**, and so on. Finally the terminal nodes are indicated, each with the value that would be predicted for an observation that fell into that node.

If you have a PostScript printer available, the function **post.tree** can be used to make a presentation plot of a **tree** object. This plot uses ellipses for nodes in the tree and rectangles for terminal nodes, and also presents additional information about the observations in each node. The command

Figure 10.4   Plots for **tree** Auto Satisfaction Example

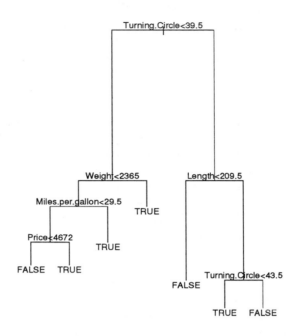

```
> post.tree(auto.reg,file="posttree.ps")
```
will produce a file called **posttree.ps**, which can then be sent to a PostScript printer to provide a representation of the tree similar to that shown in Figure 10.5.

10.11 Nonlinear Regression (nls)

The formulas for most of the other modeling functions do not require explicit information about the functional form of the modeling equation because they are limited to additive models, either as linear terms or with some parametric or smoothed function applied to a term. The class of nonlinear models specifies a functional form that relates the independent variables to the dependent variable, and then attempts to estimate the coefficients of that function that result in the lowest sum of squared residuals. This means that when you write a formula for use with **nls**, as opposed to the modeling functions discussed in the previous sections, you must explicitly include the parameters you wish to estimate in the formula. In addition, whereas both **lm** and **aov** also use the least-square criteria,

Figure 10.5   Presentation Plot for **tree** Auto Satisfaction Example

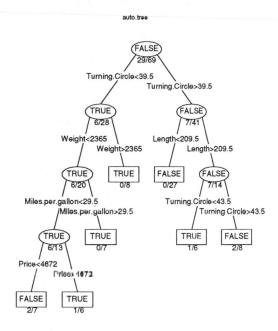

the parameter values for nonlinear models need to be estimated in an iterative fashion, starting with an initial guess that you must provide. Sometimes a poor choice of initial values may cause the algorithm to fail to converge; in other cases, no solution may be available regardless of the starting values. The initial values are passed to **nls** through the **start** argument, which consists of a list with named components that correspond to each of the parameters to be estimated. Alternatively, you can store starting values as parameters in a data frame. This technique is illustrated in the example below.

Interactions and nested terms are not permitted in **nls** models, so the special operators listed in Table 10.1 will have their usual meanings when they are used in formulas passed to **nls**.

10.11.1   Example: Nonlinear Regression

The data for this example consist of a simulated data set containing concentrations of starting material and yield of product for a hypothetical chemical reaction. The data are displayed in the table below.

Conc.	0.5	1.0	1.5	2.0	2.5	3.0	3.5	4.0	4.5	5.0
Yield	150.2	188.9	210.9	222.8	249.8	257.2	246.4	295.5	270.3	262.3

The first step is to read the data and form a data frame; a plot is then produced to examine the relationship between concentration and yield.

```
> chemdata <- data.frame(
+   Conc = c(0.5, 1.0, 1.5, 2.0, 2.5, 3.0, 3.5, 4.0, 4.5, 5.0),
+   Yield = c(150.2, 188.9, 210.9, 222.8, 249.8,
+             257.2, 246.4, 295.5, 270.3, 262.3))
> plot(chemdata$Conc,chemdata$Yield,xlab="Concentration",
+                          ylab="Yield",
+                          main="Concentration versus Yield")
```

The plot, shown in Figure 10.6, suggests a relationship of the form

$$\text{Yield} = \alpha + \beta e^{\gamma \text{Conc}}$$

Thus there are three parameters that will be estimated: α (alpha), β (beta), and γ (gamma). These parameters must be explicitly included in the formula that is passed to nls; in addition, starting values for each of the parameters must be passed to nls through the start argument. The starting values are passed in a list with one element for each parameter to be estimated. Each element should be named to correspond to the parameter it represents. For this example, suppose we use starting values of alpha = 200, beta = −20, and gamma = −0.1. The following statements perform the nonlinear regression and display the results:

```
> chem.nls <- nls(Yield ~ alpha + beta * exp(gamma * Conc),
+                 data=chemdata,
+                 start=list(alpha=200, beta=-20, gamma=-0.1))
> chem.nls
Residual sum of squares : 1282.922
parameters:
     alpha        beta       gamma
 282.0562 -180.3221 -0.6298791
formula: Yield ~ alpha + beta * exp(gamma * Conc)
10 observations
> summary(chem.nls)

Formula: Yield ~ alpha + beta * exp(gamma * Conc)

Parameters:
             Value Std. Error    t value
alpha   282.056000  15.301600   18.43320
 beta  -180.322000  22.087400   -8.16403
gamma    -0.629879   0.215205   -2.92688
```

Figure 10.6   Plot of Concentration vs. Yield for **nls** Example

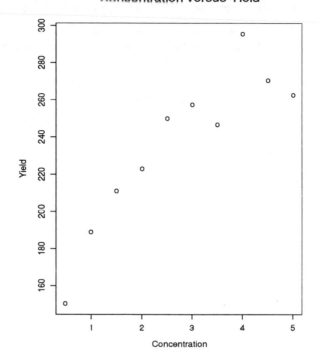

Concentration versus Yield

```
Residual standard error: 13.5379 on 7 degrees of freedom

Correlation of Parameter Estimates:
        alpha   beta
  beta 0.103
 gamma 0.900 0.446
```

The print method for **nls** objects displays the residual sum of squares at the final iteration, along with the parameter estimates and the formula. The summary method displays the parameters along with their asymptotic standard errors and asymptotic t values, and a matrix showing the asymptotic correlations among the parameter estimates. Although there is no prediction method for **nls** objects that will accept new data values, the function **predict** or **fitted** will provide fitted values for the data points. There is no plot method for **nls** objects.

As an alternative to specifying the starting parameter values in the call to **nls**, these values can be stored along with the data frame itself by using the **parameters** function. In the current example, we could store the starting values for **alpha**, **beta**, and **gamma** in the **chemdata** data frame with the following statement:

```
> parameters(chemdata) <- list(alpha=200, beta=-20, gamma=-0.1)
```

This converts **chemdata** to a parameterized data frame, or **pframe**. We could now fit the nonlinear model to the data in **chemdata** without specifying starting values, since they are stored along with the data:

```
> chem.nls <- nls(Yield ~ alpha + beta * exp(gamma * Conc),
+                     data=chemdata)
```

The print method for parameterized data frames will print the values of the parameters along with the data.

```
> chemdata
Parameters:
$alpha:
[1] 200

$beta:
[1] -20

$gamma:
[1] -0.1



Variables:
     Conc Yield
 1   0.5 150.2
 2   1.0 188.9
 3   1.5 210.9
 4   2.0 222.8
 5   2.5 249.8
 6   3.0 257.2
 7   3.5 246.4
 8   4.0 295.5
 9   4.5 270.3
10   5.0 262.3
```

10.12 Simple Statistical Inference

In S-PLUS, a number of functions are available to perform tests of statistical hypotheses. These functions are summarized in Table 10.4.

Each of the inference procedures returns an object of class **htest**, for which a printing method is available. The returned object for each of these tests will

Table 10.4 Statistical Inference Procedures

Function	Description
binom.test	Exact test for binomial proportion
chisq.test	Chi-square test for contingency table
cor.test	Test for zero correlation
fisher.test	Fisher's exact test for contingency tables
friedman.test	Friedman's rank sum test for randomized blocks
kruskal.test	Kruskal-Wallis rank sum test
mantelhaen.test	Mantel-Haenszel Chi-square Test
mcnemar.test	McNemar's Chi-square Test
prop.test	Tests for proportions
t.test	Student's t test (one-sample, two-sample, and paired)
var.test	F test to compare two variances
wilcox.test	Wilcoxon rank sum and signed rank sum tests

display information about the samples and hypotheses being tested, the test statistics and associated probability levels, and related confidence intervals. Refer to the individual help files to learn about the components that are contained in objects returned by these functions, and that can be used in more complex computations or custom displays.

Many of the inference functions allow you to specify whether the test you perform is one- or two-sided, and, for one-sided tests, whether you wish to consider only the possibility that the actual value is less or more than the hypothesized one. This is done through the `alternative` argument, which accepts values of `"two.sided"`, `"less"`, or `"greater"`.

The `binom.test` function tests whether data from a set of binomial trials come from a binomial distribution with a specified probability. It performs an exact test, not an approximation, so it is valid with all sample sizes. A binomial trial can be thought of as any independent event that may take on exactly one of two values, referred to as success and failure. Input to the function consists of x, the observed number of successes; n, the total number of trials; and p, the hypothesized probability, which defaults to 0.5 if it is not given. The function also accepts the `alternative` argument described above.

A related test is performed by the function `prop.test`, which compares proportions to hypothesized values, or tests for the equality of several proportions. The arguments to `prop.test` are x, a scalar or vector giving the number of successes; n, a scalar or vector giving the number of trials and p, a scalar or vector giving the hypothesized probability. If any of the arguments are vectors, they must all be the same length; each element of the vectors represents one proportion that is being tested. If x and n are both scalars, and no value for p is specified, a test for equality of the proportion to 0.5 is performed. In addition, the `alternative` argument, described above, is also available in `prop.test`. This function calculates a chi-square statistic, so the significance levels and confidence intervals it reports are asymptotic, not exact.

Several functions provide tests for contingency tables. (See Section 3.11 for information on producing contingency tables.) The function chisq.test performs a classical chi-square analysis, providing asymptotic tests for the independence of rows and columns in a contingency table. It accepts either a single argument, which should be a two-dimensional matrix, such as returned by table, which contains the counts of the contingency table, or two arguments, each of which should be categories or factors of the same length giving the group membership of the two factors being studied.

An exact test for a contingency table can be performed using fisher.test. It accepts its input in the same fashion as chisq.test but uses combinatorial methods to provide an exact test of the independence of rows and columns. For this reason, it will be much slower than chisq.test, and it is limited to factors or categories with less than 10 levels.

When your data consist of several 2×2 tables, recorded at various levels or strata of a third variable, the mantelhaen.test function can be used to calculate the Mantel-Haenszel chi-square test. Input to this function is either a three-way array (as produced by table), whose first two dimensions are each 2, or three vectors of categories or factors giving group membership for the observations. In the latter case, the first two arguments to mantelhaen.test must have exactly two levels.

When you have measures of a categorical variable before and after some treatment, and the levels for the two variables are the same, the mcnemar.test function can be used to perform McNemar's test of symmetry for a contingency table. This tests the hypothesis that an observation classified into category i before the treatment and into category j after the treatment is just as likely as one to be classified into category j before the treatment and into category i after the treatment. The input is either a square contingency table, as produced by table, or two vectors of categories or factors, each of which have the same levels.

Correlations between two numeric vectors can be tested for equality to zero with the function cor.test. The arguments to this function are x and y, the vectors in question, and a character argument method, which can be either "pearson", "kendall", or "spearman", depending on which type of correlation is desired. This function also accepts the alternative argument.

Three varieties of Student's t test are available through the function t.test. Each test assumes that the data arise from a normal distribution. If only one numeric vector is provided through the x argument, then, by default, t.test performs a test of the hypothesis that the mean of the vector is equal to zero. You can specify a different hypothesized value through the mu argument. If a second numeric vector is passed to t.test through the y argument, then a test that the difference of the means of x and y is equal to zero is performed; the hypothesized difference can be set to some other value using the mu argument. Finally, if the argument paired is set to TRUE, then a paired t test is performed, testing the hypothesis that the mean difference between observations is 0 (or mu). In this case, the elements of **x** and **y** should represent paired values for several observations, and **x** and **y** must be of the same length. By default, the two sample and paired tests assume that the variance of both samples is equal. If you suspect this is not the case, a modified *t* test, known as the Welch Modified *t* test, can be performed by setting the argument **var.equal** to **FALSE**. In all cases, the argument **alternative** can also be specified to control whether the test is one-sided or two-sided.

The function **var.test** uses an F test to test the hypothesis that two variances from normally distributed samples are equal. The function accepts two numeric arguments, **x** and **y**, representing the two samples whose variances are to be compared. The **alternative** argument may be supplied, as well as an argument named **conf.level**, which controls the level of the calculated confidence interval on the ratio of the two variances.

Three nonparametric counterparts to standard normal theory tests are provided. The function **wilcox.test** performs a Wilcoxon signed rank test for two independent samples (equivalent to a Mann-Whitney test), two paired samples, or a single sample. Thus, it provides a nonparametric counterpart for **t.test** and accepts the same arguments as the **t.test** function. In addition, the argument **exact** can be used to determine whether the exact distribution of the test statistic or a normal approximation is used to calculate the probabilities reported by the function. The default is **exact=TRUE** if the sample size is less than 49, and **exact=FALSE** otherwise.

The function **kruskal.test** performs a Kruskal-Wallis rank sum test for a one-way analysis of variance. It requires two arguments, each of the same length, with corresponding elements representing information for a single observation. The argument **y** consists of the dependent variable values, whereas **groups** is a factor or category giving group membership. For an unreplicated two-way randomized block design with no replicates, **friedman.test** performs a rank sum test similar to **kruskal.test**. Along with the two arguments of that function, a third, **blocks**, is required and should be a factor or category containing the block information, in a fashion similar to the **y** and **groups** arguments.

10.13   Multivariate Statistical Procedures

In addition to the statistical modeling and hypothesis testing procedures described in previous sections of this chapter, S also offers a number of functions for performing multivariate statistical procedures. Since these statistical techniques do not fit models or test hypotheses, they are not as well unified into S as some of the techniques described previously, but they are still very useful for working with data sets that have many variables that need to be considered simultaneously. Table 10.5 lists the functions that are available, along with a brief description of each.

Table 10.5 Multivariate Statistical Procedures

Function	Description
cancor	Canonical correlation
cmdscale	Multidimensional scaling
discr	Linear discriminant analysis
dist	Calculate a distance matrix
hclust	Hierarchical clustering
clorder	Reorder a cluster tree
cutree	Form groups from a cluster analysis
labclust	Label a cluster plot
leaps	All subsets regression using leaps and bounds
plclust	Produce a cluster plot (dendogram)
subtree	Extract part of a cluster tree
prcomp	Principal components analysis

Canonical correlation is useful when your variables can be divided logically into two groups; the technique finds linear combinations of variables in each of the groups to maximize the correlation between the two linear combinations. The cancor function requires two arguments, which are matrices containing the variables in each of the two groups, and returns a list with the following elements: cor, the correlations between the variables defined by the linear combinations; xcoef and ycoef, which contain the coefficients of the linear combinations; and xcenter and ycenter, which contain values subtracted from each column of the input matrices. (By default, cancor subtracts the column mean for each variable from its input arguments.)

The cmdscale function for multidimensional scaling takes as input an $n \times n$ distance matrix, such as that produced by dist, representing the distances among a set of n observations, and attempts to find a set of k dimensions that explain the relative distances among the observations. It accepts as input the distance matrix and k, the number of dimensions desired, and returns an $n \times k$ matrix, each row of which contains the coordinates in the k-dimensional space of an individual observation.

Linear discriminant analysis is useful in the case in which p variables are available for each of several cases that can be classified into k known groups. The discr function expects its data (which is the first argument to the function) to be a matrix of explanatory variables whose rows are ordered by group membership, and requires a second argument that is either a scalar, giving the number of groups if each has the same number of observations, or a vector of length equal to the number of groups, containing the number of observations in each of the groups. Suppose the matrix x is a matrix containing the explanatory variables you wish to use in the discriminant analysis, and group is a vector of the corresponding group membership. You can arrange your data for input to the discr function with statements such as the following:

```
> size <- table(category(sort(group)))
> x.discr <- discr(x[order(group),],size)
```

The output from `discr` is a list containing three components: `cor`, a vector of correlations showing the strength of the linear combinations that are used to discriminate among the groups; `var`, the matrix of linear combinations that are applied to the variables to create the discriminant variables; and `groups`, which provides linear combinations of group contrasts that maximize the correlations with the discriminant variables.

The `dist` function accepts as input an $n \times m$ matrix of data, and returns a structure that contains the distances between each of the n observations represented by the matrix. An optional second argument defines the distance metric to be used. Choices include `"euclidean"`, `"maximum"`, `"manhattan"`, and `"binary"`. Since a distance matrix is by definition symmetric, and each diagonal element is equal to zero, only the lower off-diagonal elements are stored in the output from `dist`, and only functions like `cmdscale` and `hclust`, which expect distance matrices as input, will use the output from `dist` appropriately. However, if you need the actual distance matrix itself, you can use statements like the following to produce it from the output of `dist`:

```
> x.d <- dist(x)
> x.dist <- matrix(0,n,n)
# fill in the lower triangle of x.dist
> x.dist[row(x.dist) > col(x.dist)] <- x.d
# set the upper triangle equal to the lower triangle
> x.dist <- x.dist + t(x.dist)
```

Hierarchical cluster analysis is useful when you have measurements of a variety of variables for a set of observations and wish to determine if there are any natural groupings of the data wherein observations in the same group (called a cluster) tend to be closer together than observations in different groups. The `hclust` function implements this technique and accepts as input a distance matrix as produced by `dist` and a character string specifying the cluster method to be used. Choices include `"average"`, `"connected"` (also known as single linkage), and `"compact"` (also known as complete linkage). It returns a structure suitable for use with several other functions: `plclust` and `labclust`, respectively, plot and label a cluster tree to the active graphics device; `clorder` and `subtree` manipulate parts of a cluster tree; and `cutree` produces a vector of group memberships based on a cluster analysis. To use `cutree`, two arguments are required. The first is the list output from `hclust`, and the second is either `k`, the number of groups desired, or `h`, the height at which the tree should be cut to create the groups. The appropriate heights are stored in the output from `dist` and are displayed on the cluster tree. The output from `cutree` can be used with functions like `tapply` or `split` to calculate statistics for the different groups in order to help define what makes the groups discerned by the cluster analysis different.

The **leaps** function uses an efficient algorithm to carry out regressions for all possible subsets of a set of independent variables. A matrix whose columns contain the candidate independent variables is given to **leaps** as its first argument, and a vector containing the dependent variable is given as the second argument. An optional argument, **wt=**, allows a weighting variable to be used in the equations. You can specify the criteria to be used to evaluate the subscripts through the **method=** argument, which must be one of "Cp", "r2", or "adjr2" corresponding to Mallows' Cp statistic, the usual squared multiple correlation, or the adjusted squared multiple correlation, respectively. If no **method=** argument is used, the Cp statistic is used to evaluate the subsets. The output consists of a list with four components. The first component is the chosen statistic, named either **Cp**, **r2**, or **adjr2**; the second (named **size**) is the size of the subset in question; the third (**names**) is a vector of character strings giving the names of the variable in the subset; and the fourth is a logical matrix called **which**, with one row for each subset, and whose columns represent the candidate variables, with values of **TRUE** corresponding to variables included in that particular subset, and **FALSE** otherwise. Thus, rows of **which** can be used as subsetting subscripts for the data matrix, allowing variables in any subset to be easily accessed for further study. There is a built-in limit of 30 independent variables for the **leaps** function.

10.14   Function Minimization (ms)

The function **ms** performs minimization of an arbitrary function using a technique known as minimum sums. As with **nls**, formulas passed to **ms** must explicitly include the parameters that you are trying to estimate; unlike any of the other modeling functions, formulas for **ms** will not have a left-hand side. However, for the S interpreter to understand that they are formulas, they must begin with a tilde (˜). You must provide starting values for **ms** in the same fashion as for **nls**, namely through an argument called **start**, which is a list of named components, one for each parameter, containing the starting values, or by storing the parameters in the data frame.

10.14.1   Example: Minimization

As an example of function minimization using **ms**, consider maximum-likelihood estimation of the parameters of the beta distribution. In maximum likelihood, the probability density for a given distribution is viewed as a function of the parameters that define the distribution, and the product of the likelihoods for all of the observations is maximized. In practice, the maximization is actually carried out on the log of the likelihood, which is the sum of the logs of the likelihoods for the individual observations. The likelihood for the beta distribution can be calculated using the function **pbeta**. In the following example, the function **rbeta** is used to generate 100 random beta distributed variables with parameters 4 and 2. If you use these statements on your computer, you will get somewhat different answers because of the nature of the random number generator. Note that, since the problem at hand is maximization, and **ms** actually performs minimization, the negative of the sum of log likelihoods is the function that is passed to **ms**.

```
> bdata <- rbeta(100,4,2)
> z<-ms(~-sum(log(dbeta(bdata,shape1,shape2))),
+              start=list(shape1=2,shape2=2))
> z
value: -30.0597
parameters:
   shape1    shape2
 3.390245 1.769401
formula:    ~   - sum(log(dbeta(bdata, shape1, shape2)))
1 observations
call: ms(formula =   ~   - sum(log(dbeta(bdata, shape1, shape2))),
        start = list(shape1 = 2, shape2 = 2))
> summary(z)
Final value: -30.0597

Solution:
           Par.
shape1 3.390245
shape2 1.769401

Convergence: RELATIVE FUNCTION CONVERGENCE.

Computations done:
 Iterations Function Gradient
          8       11       20
```

The print method for **ms** displays the value of the function at the minimum (in this case -30.0597), as well as the parameter values. The summary method gives additional information about the progress of the iterative process. No plot method is available for **ms** objects.

Exercises

1: In experiments in which the dependent variable consists of counts, an analysis of variance with a square root transformation is often recommended. An alternative is to fit a generalized linear model, using a **poisson** link function. The **solder2** data set from the **data** library has information about the number of solder skips in circuit boards as well as five experimental factors that may affect the number of skips. Using **skips** as a dependent variable, fit both the square root and the Poisson model, using main effects only, as well as second-order interactions. Which seems to do a better job of describing the data? (Since the models use different transformations, studying the relationships of the fitted values to the actual values may be helpful.)

2. The data set **Lubricant** in the **data** library gives the viscosity (**viscos**) of a lubricant measured at different pressures (**pressure**) at each of four different temperatures (**tempC**). First, make a plot of the relationship between viscosity and pressure using a different line for each of the temperatures. Do the lines look parallel? Test for parallelism by fitting a linear model with terms for **pressure**, **tempC**, and **tempC:pressure**. Does the model support your visual assessment? (Make sure to convert **tempC** to a factor before performing the analysis.)

3. The data set **ethanol** in the **data** library reports on an experiment involving automobile emissions. The variables in the data set are **NOx**, the amount of nitric oxide emitted; **E**, the equivalence ratio, which measures the ratio of fuel to air; and **C**, the compression ratio of the engine. Plot **NOx** versus **E** to get a feel for the relationship between these two variables. Using the **bs** function, fit a B-spline on **E** to predict **NOx**, using **gam**. What kind of results would you expect from a linear regression with this data?

11 Applied Statistical Models

11.1 Graphics and Statistical Models

Although the plot methods for specific modeling techniques provide a first look at the results of statistical models, there are a variety of other graphics functions that are useful in evaluating how well a statistical model fits a particular data set, as well as in testing the assumptions on which the statistical methods are based. For example, for probabilities associated with some statistical tests to be valid, the unexplained error in the data should follow a normal distribution. Thus, testing the residuals of an analysis to see if they follow a normal distribution is often a useful adjunct to an analysis. The function `qqnorm` plots the quantiles of a vector against the quantiles of a standard normal distribution; if the vector in question has a normal distribution, this plot should follow a straight line. The function `qqline` will draw this line for visual comparison. For example, suppose we carry out a linear regression on the reciprocal of `Miles.per.gallon` versus `Weight` for the `auto` data frame described in Section 10.1. (For more details about the `lm` function, see Section 10.5.) We could perform the regression and produce a normal quantile plot of residuals with the following statements:

```
> rmpg.reg <- lm(1/Miles.per.gallon~Weight,data=auto)
> qqnorm(residuals(rmpg.reg))
> qqline(residuals(rmpg.reg))
```

The plot is shown in Figure 11.1. Except for the three most negative values, the residuals appear to be following a normal distribution. These three values could be investigated to determine whether they provide insight into the discrepancy.

One type of plot that is useful for gaining insight into the relationship between variables is the conditioning plot, which is implemented in S through the function `coplot`. The basic idea behind a conditioning plot is to consider the relationship between two variables of interest while holding one or two other variables fixed at a set of values that spans their range. The variables that are held fixed are known as given variables; their values are displayed at the top

Figure 11.1   Normal Quantile Plot of Residuals

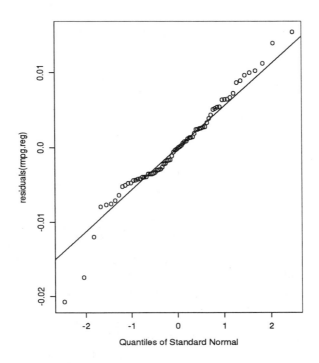

of the plot and, if two given variables are used, at the side of the plot. The portions of the plot in which the given values are displayed are known as the given panels. The relationship between the variables of interest is shown in the remaining part of the graphical display, known as the dependency panels. By default, these panels contain scatterplots of the data of interest for observations that have specific values of the given variables. (You can use the optional **panel** argument of **coplot** to perform some other action.) The **coplot** function accepts a formula relating the two variables of interest, and uses the vertical bar (|) to separate the formula from the specification of the given variables. For example, suppose we wish to examine the relationship between fuel economy and weight for automobiles, using displacement as a given variable. The statement

```
> coplot(Miles.per.gallon ~ Weight | Displacement, data=auto)
```

produces the graph shown in Figure 11.2. In the dependency panel, the dotted line represents the switch between the lower and the upper line of plots in the dependency panel; the first bar from the left, centered around 100, indicates that the lower left-hand plot is for those points where **Displacement** ranges from about 80 to 120; the next plot has points where the range of **Displacement** values is about 100 to 150, and so on. Thus, from the coplot in Figure 11.2 it can

Figure 11.2   Coplot of Miles per Gallon vs. Weight with Displacement Given

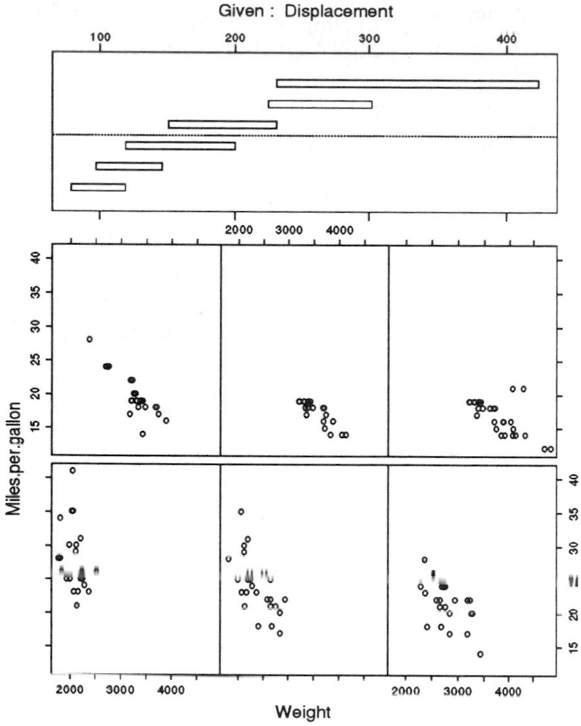

be seen that the relationship between **Weight** and **Miles.per.gallon** is much stronger for lower values of **Displacement** than for higher values.

You can specify the number of rows and columns to be used in the dependency panels through the arguments **rows** and **columns** in the **coplot** function. To specify the ranges to be used in the given panel, the argument **given** can be used in **coplot**. This argument should be either a vector of discrete values or a two-column matrix whose first column contains starting values and whose second column contains ending values for each of the ranges. The function **co.intervals**, used internally by **coplot**, is useful in this connection, because it allows you to specify the number of intervals produced as well as the fraction of points that will be shared by successive intervals, and produces output that is suitable for use with the **given** argument of **coplot**.

For the case of two given variables, **coplot** will produce an array of plots in the dependency panel, with given panels both above and to the right of the array. The **rows** and **columns** arguments are ignored in this case. As an example, consider the **state** data frame, created from the data matrix **state.x77**. A scatterplot of life expectancy (**Life.Exp**) versus illiteracy (**Illiteracy**) shown in Figure 11.3, shows that the populations of states with higher illiteracy rates appear to have lower life expectancies. To view the effects of other variables, for example **Income** and **Area**, we could produce a coplot, using these two variables as givens. Because there are not many data points, we use the **given** argument to specify four levels for each of the given variables, resulting in a 4-by-4 array of plots in the dependency panel. Note that with two given variables, the **given** values are passed to **coplot** through a list. The coplot itself is displayed in Figure 11.4.

```
> state <- data.frame(state.x77)
> plot(state$Illiteracy, state$Life.Exp,
+      main="Plot of Life Expectancy vs. Illiteracy",
+      xlab="Illiteracy Rate",ylab="Life Expectancy")
> coplot(Illiteracy~Life.Exp|Income*Area, data=state,
+        given=list(co.intervals(state$Income,4),
+                   co.intervals(state$Area,4)))
```

Figure 11.3   Scatterplot of Life Expectancy versus Illiteracy

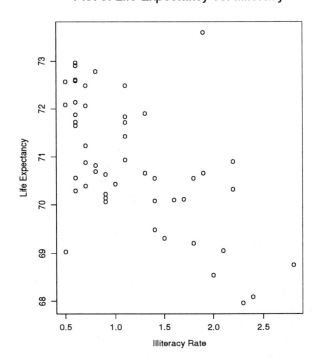

The coplot shows that the relationship is strongest in those states with lower average income levels; the plots on the left-hand side of the dependency panel, corresponding to lower income levels, show the most dramatic downward trends.

Figure 11.4   Coplot with Two Given Variables

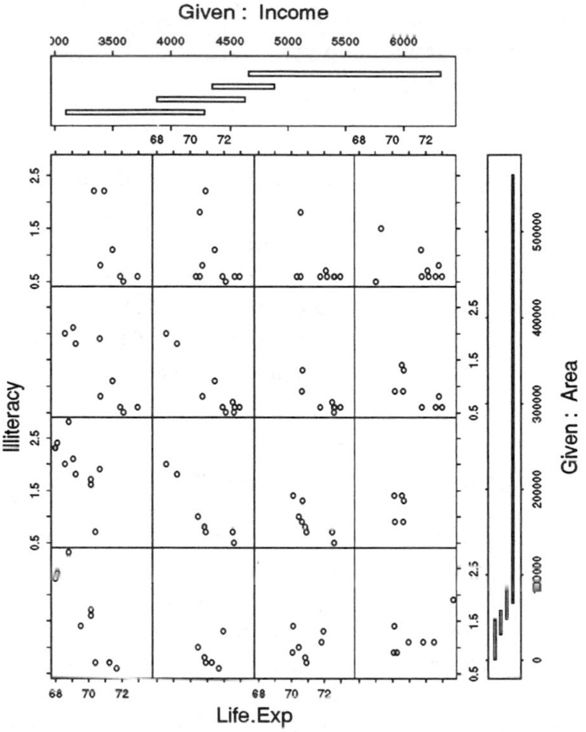

11.2   Sequential and Partial Sums of Squares

The default analysis displayed by the summary method calculates sums of squares known as sequential sums of squares, because each effect in the table is evaluated after those effects preceding it have been evaluated. If the design is balanced— that is, if the number of observations for each combination of factors in the formula is the same—this ordering makes no difference in the calculation of sums of squares and associated hypothesis tests. However, in an analysis such as this one, which is unbalanced, an alternative method of calculating sums of squares is often used. This technique, known as partial sums of squares, calculates the sums of squares for each effect in the model as if it were the last one added. The function **drop1** can be used to print an ANOVA table containing partial sums of squares. The first argument to **drop1** is an object resulting from an **lm** or **aov** analysis, or from some other method that inherits from those classes. The optional argument **scope** is used to pass a formula containing those terms which should be treated in turn as the last term in the model. Only the right-hand side of the **scope** formula is used, and, in this context, a period (.) means all the terms that were included in the formula of the object passed to **drop1**. For the `auto.aov2` analysis, a table displaying partial sums of squares can be generated with the following statement:

```
> drop1(auto.aov2,scope=~.)
Single term deletions

Model:
Miles.per.gallon ~ Man + Cgrp + Man:Cgrp
          Df Sum of Sq      RSS  F Value     Pr(F)
  <none>                118.2500
     Man  3    7.6363 125.8863  0.38747 0.7633766
    Cgrp  1  102.2177 220.4677 15.55957 0.0009502
Man:Cgrp  3   21.7152 139.9652  1.10183 0.3740998
```

The line labeled `<none>` corresponds to the residual sum of squares.

11.3   Contrasts and Model Matrices

As mentioned in Section 10.6, to overcome the problem of overparameterization in creating a model matrix, the **aov** function generates a set of contrasts that are estimated, instead of trying to estimate a separate parameter for each level of a factor in the model. For unordered factors, the default choice of contrasts is a Helmert contrast. This form of contrast compares the second through last levels of a factor with the mean of previous levels, resulting in one contrast less than the number of levels of the factor. For ordered factors, the default is orthogonal polynomial contrasts. There will be one less contrast than the number of levels; each one will correspond to a polynomial. The first contrast represents a linear polynomial (straight line), the next is a quadratic polynomial, and so on. These contrasts are generated by the functions `contr.helmert` and `contr.poly`, respectively. The default choice of contrasts can be changed through the S option `contrasts`.

As an example of a polynomial contrast, consider the `Cargrp` factor, designated as an ordered factor in Section 10.1. We can view the contrasts that will be used in building the model matrix with the function `contrasts`:

```
> contrasts(auto$Cargrp)
               .L          .Q
  Small -0.7071068  0.4082483
 Medium  0.0000000 -0.8164966
  Large  0.7071068  0.4082483
```

Because there are three groups, whose labels are shown as row names in the output from `contrast`, two contrasts are constructed: a linear one and a quadratic one, with column labels `.L` and `.Q`, respectively. These contrasts assume that the spacing between the ordered levels is equal; if it is not, the function `contr.poly` can be called directly to construct appropriate contrasts. To actually construct the part of the model matrix corresponding to the `Cargrp` factor, this contrast matrix is used to multiply a dummy matrix, consisting of one row per observation and as many columns as there are levels of the factor. Each entry of the dummy matrix is set to zero, except for exactly one element of each row corresponding to the level of the `Cargrp` for that observation, which is set to one. The function `model.matrix` does the actual work of producing the model matrix, and can be called directly with a formula if you need access to it. The portion of the model matrix corresponding to interactions is constructed from the corresponding rows of the model matrix for the main effects contained in the interaction.

In many analyses, the overall question of whether or not the mean differs among all the levels of a factor is not as interesting as are some specific comparisons. For example, several fertilizers may be tested for their effectiveness in increasing the yield of a crop, but the main comparison of interest might be that between just a single new fertilizer and its leading competitor. It is still of great value to include the other levels of the factor in the analysis, because they help to increase the precision with which the residual error is estimated, leading to more sensitivity in detecting differences. To test comparisons like this, you can temporarily specify a custom contrast matrix to be used to construct a model matrix using the `C` function. Then the `proj` function can be used to calculate the projection of the dependent variable on the orthogonalized model matrix. The matrix that is output from `proj` decomposes the dependent variable into orthogonal components, which, for each row, sum up to the value of the dependent variable for the corresponding observation. In addition, the sum of squares due to the contrast will be the sum of the squared elements in the corresponding column of the `proj` output matrix. Consider a factor with k levels. If you specify fewer than $k - 1$ contrasts in the matrix you pass to `C`, it will automatically create additional contrasts, orthogonal to the ones you created, to complete the contrast matrix. Note that the overall sums of squares for the effects will not be changed when you specify your own contrasts, provided that they are not linearly dependent. If they are linearly dependent, the number of degrees of freedom for the effect will be reduced or the `aov` function will report singularities in the model.

As an example of a custom contrast matrix, consider the one-way ANOVA presented in Section 10.6.1. Suppose we are interested in comparing the mean fuel economy of Buicks against that of Chevrolets. The proper ordering for specifying the contrast is determined by the `levels` attribute of the factor in question. Each row of the contrast matrix corresponds to one of the levels of the factors; each column represents a single-degree-of-freedom contrast. The contrast matrix to compare the mean fuel economies for the two manufacturers can be constructed as follows:

```
> levels(auto.use$Man)
[1] "Buic" "Chev" "Olds" "Pont"
> my.contrast <- matrix(c(1,-1,0,0),ncol=1)
> dimnames(my.contrast)<- list(levels(auto.use$Man),
+                                    "Buick vs. Chevy")
```

```
> contr.aov <- aov(Miles.per.gallon~C(Man,my.contrast),
+                   data=auto.use)
> contr.proj <- proj(contr.aov,onedf=T)
> ssq <- apply(contr.proj^2,2,sum)
> my.ssq <- ssq[grep("Buick vs. Chevy",names(ssq))]
> f.con <- my.ssq / (sum(residuals(contr.aov)^2) /
+                   contr.aov$df.residual)
> c(MS=my.ssq,F=f.con,prob=1-pf(f.con,1,contr.aov$df.residual))
      MS        F       prob
 24.89032 2.300096 0.1342902
```

Even though only one column of the contrast is specified, the **contrasts** function fills in the other columns with contrasts that are orthogonal to the one specified, and to each other. (This action can be suppressed with the optional **how.many** argument.) The **onedf=T** argument to **proj** specifies that we wish to see the projection of each single-degree-of-freedom comparison, which would normally be pooled to evaluate the overall effect of the factor in question. The sum of squares corresponding to each of these contrasts is obtained by taking the sum of the squares of all the elements in the appropriate column. This sum of squares is divided by the error mean square to form an F ratio, and the results are combined in a vector for display. Note that the error sum of squares was calculated as the sum of the squared residuals from the analysis.

Sums of squares due to contrasts obtained in this way are sequential sums of squares, not partial sums of squares. (See the preceding section for a discussion of the differences.) To get the partial sums of squares corresponding to a contrast, you can fit a regression to a model matrix that contains the desired contrast as one of its columns, and use the **drop1** function to treat the contrast in question as if it were the last regressor added to the model. The function **model.matrix** can be used to generate the model matrix. Note that the model to be fitted is specified without an intercept (**-1** in the model formula), because the model matrix contains a column of 1s representing the intercept. Continuing the previous example, we can construct the model matrix from the **aov** object (**contr.aov**) and create an appropriate data frame to regress the individual contrasts against the dependent variable. Since the **model.matrix** function returns an object that will be treated as a matrix by the modeling functions, the function **matrix** is used to convert it into a regular matrix:

```
> contr.mat <- model.matrix(contr.aov)
> c.frame <- data.frame(matrix(contr.mat,ncol=ncol(contr.mat)))
> names(c.frame) <- c("Intercept","Buick.v.Chevy","Con2","Con3")
> c.frame$Miles.per.gallon <- auto.use$Miles.per.gallon
> contr.reg <- lm(Miles.per.gallon ~ . - 1,data=c.frame)
```

```
> drop1(contr.reg)
Single term deletions

Model:
Miles.per.gallon ~ Intercept + Buick.v.Chevy + Con2 + Con3 - 1
               Df Sum of Sq      RSS        Cp
       <none>                  238.07    324.64
    Intercept  1  10356.93  10595.00  10659.93
 Buick.v.Chevy  1     26.37    264.45    329.37
         Con2  1      4.58    242.65    307.58
         Con3  1      3.42    241.49    306.42
```

The output from **drop1** provides the sum-of-squares value of 26.37; the F-test
and probability could be calculated in a fashion similar to that of the preceding
example.

11.4   Regression Diagnostics

The plot methods for **lm**, **aov**, and **glm** all display residuals plotted against fitted
values in order to help identify outliers, which are individual observations that
are different in some way from most of the observations in a data set. A wide
variety of other measures, often known as regression diagnostics, can also be used
to identify outliers. Rather than attempt to include every regression diagnostic,
S provides functions that compute the quantities required for most commonly
used diagnostics. These quantities, and the functions which produce them, are
summarized in Table 11.1.

Table 11.1 Quantities Useful for Diagnostics

Symbol	Description	Function	Name of Element
h	Diagonal of hat matrix	lm.influence	hat
b or $\hat{\beta}$	Coefficients	use coef()	
r or e	Residuals	use residuals()	
s or $\hat{\sigma}$	Residual standard error	summary	sigma
$s_{(i)}$	Leave out one estimate of $\hat{\sigma}$	lm.influence	sigma
$b_{(i)}$	Leave out one estimate of b	lm.influence	coefficients
$(X'X)^{-1}$	Unscaled covariance of b	summary	cov.unscaled

The vector **sigma** returned by the **lm.influence** function contains estimates of the residual standard error of the model calculated by excluding one
observation at a time. Thus, the length of the vector will be equal to the number
of observations in the data set that created the analysis object passed to the function. In a similar fashion, the matrix **coefficients** returned by **lm.influence**
is a matrix with as many rows as there are observations in the data set, where

each row contains the estimates of the coefficients of the model calculated by excluding one observation at a time.

11.5   Predicted and Fitted Values

One of the main goals of statistical modeling is to come up with a means of predicting dependent-variable values, given a set of independent-variable values. Oftentimes, before a model is applied to a new set of data, it is of interest to examine how well the model performs in predicting the known values of the observations that were used to fit the model. The function `predict` will provide these values when it is passed a single argument representing an analysis object from `lm`, `aov`, `glm`, `gam`, `loess`, `tree`, or `nls`. Keep in mind that for models with link functions other than the identity, the values returned by `predict` are predictions for the linear part of the model, and will not be in the same scale as the dependent variable; in such cases, the function `fitted` should be used to get predictions that are in the same scale as the dependent variable.

To predict values for observations that were not used to fit the model, the optional argument `newdata` can be used to pass a data frame containing independent-variable values to the `predict` function. (This argument is not available with `nls` objects.) The data frame passed as a `newdata` argument must have variables with the same names as those that were used in the original model. For example, consider the `auto.reg` object produced in Section 10.5.1. This was the result of a linear regression using the formula `Miles.per.gallon ~Length + Weight + Displacement`. Suppose we wish to use this model to predict `Miles.per.gallon` for a car with the following values of the independent variables in the model:

```
Length Weight Displacement
   190   3000          170
```

To use the `predict` function with new data, we need to pass them in the form of a data frame:

```
> predict(auto.reg,newdata=data.frame(Length=190,
+          Weight=3000,Displacement=170))
         1
 20.96847
```

In some models, transformations are carried out that are data-dependent: when a new observation is presented for prediction, it may not be possible to directly substitute in the fitted model to get an answer. This situation arises most commonly in **gam** models, because the core of the technique is to find data-dependent transformations for the independent variables that are effective in estimating the dependent variables. Thus, the prediction method for **gam** models, incorporated into the function `predict.gam`, has special provisions for handling this situation. Considering the example in Section 10.8.1, where we fitted the model `Loss ~s(AirFlow) + s(WaterTemp) + s(AcidConc)`, if we create a data frame containing two new observations of air flow, water temperature, and acid concentration, we can get predicted values directly through a simple call to predict:

```
> new.data <- data.frame(AirFlow=c(55,62), WaterTemp=c(18,24),
+                        AcidConc=c(85,87))
> predict(stack.gam,newdata=new.data)
       1        2
11.57808 24.04146
```

However, **gam** models are not the only ones in which data-dependent transformations may be carried out. The function **poly**, for example, will generate a matrix of orthogonal polynomial components of a data set, suitable for use in polynomial regression. Suppose we wish to fit a polynomial regression model to the **Loss** variable in the **stack** data frame, using **WaterTemp** as an independent variable. We can fit a third-order polynomial regression with the statements

```
> stack.poly<-lm(Loss~poly(WaterTemp,3),data=stack)
> summary(stack.poly)
```

```
Call: lm(formula = Loss ~ poly(WaterTemp, 3), data = stack)
Residuals:
    Min      1Q Median     3Q    Max
 -4.877  -2.891 0.1091  3.109  8.108
```

```
Coefficients:
                     Value Std. Error  t value
        (Intercept)  17.524    0.8428  20.7931
poly(WaterTemp, 3)1  39.826    3.8621  10.3120
poly(WaterTemp, 3)2  15.020    3.8621   3.8891
poly(WaterTemp, 3)3  -1.996    3.8621  -0.5167
                        . . .
```

Now suppose we wish to obtain the fitted value for an observation with water temperature equal to 25. The usual **predict** function (actually **predict.lm**) cannot handle this situation, because it will try to reconstruct the orthogonal polynomial components for the new data, which will not be possible for a single point. By calling **predict.gam** directly, however, predictions can be properly calculated:

```
> predict.gam(stack.poly,newdata=data.frame(WaterTemp=25))
[1] 28.89248
```

For tree-based regression models, the **predict** function works in a similar fashion to the predict method for **lm** objects. However, for classification models, **predict** will return a matrix with as many rows as there are observations, and as many columns as there are levels of the dependent variable. Each row of the output from **predict** will represent the predicted probabilities of the observation falling into the different categories. For example, the **auto.tree** object produced in Section 10.10.1 would produce the following values if it were passed to `predict`:

```
> a.predict <- predict(auto.tree)
> a.predict
                      FALSE       TRUE
     Amc Concord 1.0000000 0.0000000
       Amc Pacer 1.0000000 0.0000000
       Audi 5000 0.0000000 1.0000000
        Audi Fox 0.1666667 0.8333333
        BMW 320i 0.0000000 1.0000000
   Buick Century 1.0000000 0.0000000
   Buick Electra 0.1666667 0.8333333
               . . .
```

An object would generally be classified into the group for which it had the highest probability; therefore, we can use the `apply` function to determine which group would be chosen for each observation using the tree-based rule. The following statements determine the classification chosen by the `auto.tree` model, and produce a contingency table showing the true and predicted classifications:

```
> Satis.predict <- apply(a.predict,1,
+                        function(x)seq(along=x)[x == max(x)][1])
> Satis.predict <- dimnames(a.predict)[[2]][Satis.predict]
> table(Satis.predict, auto.tree$y)
        FALSE TRUE
 FALSE    38    4
  TRUE     2   25
```

The subscript of 1 applied at the end of the function definition in the `apply` call insures that only a single value will be chosen in case of ties. The second statement provides appropriate labels for the predicted values. Finally, the true values of `Satis` are extracted from the `auto.tree` object instead of from the `auto` data frame because some observations were eliminated from the analysis because of missing values.

11.6   Updating and Comparing Models

In the course of building a statistical model, it is often useful to use a particular model as a starting point, and to modify that model by adding or deleting terms, removing observations, or even testing the model on a different set of data. To avoid unnecessary respecification of information at each of these steps, the function `update` allows you to specify only those changes that you specify when you wish to recalculate a model. The first argument to `update` is always the model object being updated; additional arguments can consist of any arguments that would be acceptable to the modeling function that originally produced the model object. In addition, you can update model formulas in an especially simple way. When updating a formula, the period (.) can be used on either side of the tilde (˜) to represent whatever was originally specified in the formula of the model object. For example, in Section 10.7.3, a log-linear model was fitted to variables in the `mtable` data frame using the following formula:

```
> formula(mtable.ll)
Counts ~ Age * Card * Income
```

It was found that the three-way interaction among `Age`, `Card`, and `Income` was not significant. To refit the model, and to eliminate this interaction, the following statement can be used:

```
> mtable.ll.new <- update(mtable.ll, . ~ . - Age:Card:Income)
> anova(mtable.ll.new,test="Chisq")
Analysis of Deviance Table
Poisson model

Response: Counts


Terms added sequentially (first to last)
```

	Df	Deviance	Resid. Df	Resid. Dev	Pr(Chi)
NULL			83	559.5742	
Age	5	93.5863	78	465.9879	0.0000000
Card	1	142.1609	77	323.8270	0.0000000
Income	6	150.0609	71	173.7662	0.0000000
Age:Card	5	3.9013	66	169.8649	0.5637172
Age:Income	30	114.1038	36	55.7612	0.0000000
Card:Income	6	30.8138	30	24.9474	0.0000275

The three-way interaction is no longer present.

The `update` function is not limited to modifying formulas. For example, in `loess` models, the argument `parametric` can be used to pass a vector of variable names that will be fitted parametrically instead of through local regressions. To update the object `stack.loess`, created in Section 10.9.1, by fitting `WaterTemp` parametrically, we could use the following statements:

```
> stack.loess.new <- update(stack.loess,parametric="WaterTemp")
> stack.loess.new
Call:
loess(formula = Loss ~ AirFlow * WaterTemp, data = stack,
        parametric = "WaterTemp")

Number of Observations:          21
Equivalent Number of Parameters: 8.4
Residual Standard Error:         9.149
Multiple R-squared:              0.69
Residuals:
    min   1st Q    median 3rd Q    max
 -3.187 -0.4794 -0.006394 1.279 22.79
```

Note that the residual standard error rose from its original value of 3.411 to 9.149. To test the significance of this difference, the function **anova** can be used. When passed more than one model object, **anova** will, when appropriate, perform a test or tests comparing those model objects. For example, we could compare the parametric version of **stack.loess** with the original version using the following statement:

```
> anova(stack.loess,stack.loess.new)
Model 1:
loess(formula = Loss ~ AirFlow * WaterTemp, data = stack)
Model 2:
loess(formula = Loss ~ AirFlow * WaterTemp, data = stack,
        parametric = "WaterTemp")
Analysis of Variance Table
        ENP     RSS     Test     F Value     Pr(F)
1       8.3   112.84   1 vs 2              54.41 0.0016032
2       8.4   799.12
```

As would be expected from the rise in residual standard errors, the formal test indicates that model 1 (the original) performed significantly better than model 2 (with **WaterTemp** fitted parametrically).

The **update** function can also be used to perform analyses on subsets of a data frame by using the **subset** or **data** arguments, which are common to all the modeling functions. To fit the regression model **auto.reg**, from Section 10.5.1, to those cars for which **Gear.Ratio** is greater than the median value, the following statements could be used:

```
> auto.reg.1 <- update(auto.reg,
+                       subset=Gear.Ratio > median(Gear.Ratio))
> auto.reg.1
Call:
lm(formula = Miles.per.gallon ~ Length + Weight + Displacement,
        data = auto, subset = Gear.Ratio > median(Gear.Ratio))

Coefficients:
  (Intercept)      Length       Weight Displacement
     46.95544 -0.01084168 -0.009547386    0.02147044

Degrees of freedom: 37 total; 33 residual
Residual standard error: 4.223825
```

The **update** function knew it should evaluate the **subset** expression in the **auto** data frame, because that was the data frame from which the original model was derived. Note that if you make changes to a data frame from which a model is derived, **update** will use the modified data set in calculating a new model; as a rule, the actual data values are not stored with the model object.

You can use **update** to fit a model to a new data frame, provided that it has the same variable names as the original data set from which the model was formed. So to fit the `auto.reg` data set to a random sample of 20 cars from the original auto data frame, the following statements could be used:

```
> new.auto <- auto[sample(nrow(auto),20,replace=F),]
> new.reg <- update(auto.reg,data=new.auto)
> new.reg
Call:
lm(formula = Miles.per.gallon ~ Length + Weight + Displacement,
        data = new.auto)

Coefficients:
 (Intercept)    Length        Weight Displacement
    67.58482 -0.215654 -0.0009130165 -0.007714412

Degrees of freedom: 20 total; 16 residual
Residual standard error: 3.655327
```

Remember that the `update` function does not try to use any previously calculated quantities when updating models, so it may not be the most efficient means of performing repeated calculations on model objects. However, its simplicity and convenience certainly make it a logical starting point for any such calculations.

11.7 Stepwise Model Selection

When you are confronted with a large number of potential independent variables in a model, it is helpful to have an automatic technique to eliminate those variables that appear to be making the least contribution to the predictive or explanatory power of the model. The function `step` uses a criterion known as AIC (Akaike Information Criterion) to determine which variables are contributing the least to the efficacy of the model at each step of the selection process. The AIC statistic starts with the residual deviance or the standard error of the model, but, since models with many terms will have smaller deviances than more parsimonious models, even if the terms are not really making a contribution, it adds a penalty to the residual deviance, which increases with the number of terms included in the model. In this way, it tries to strike a balance between reducing residual deviance and creating a model with not too many terms.

As an example, a logistic regression model was fitted to the `Satis` variable of the `auto` data frame in Section 10.7.2, using the following formula:

```
> formula(auto.lreg)
Satis ~ Price + Miles.per.gallon + Headroom + Rear.Seat + Trunk +
        Weight + Length + Turning.Circle + Displacement +
        Gear.Ratio + Cargrp
```

The analysis-of-deviance table showed that not all of the terms appeared to be contributing significantly to the model. The function `step` will delete variables one at a time until further deletions no longer produce a sizable reduction in AIC. Alternatively, certain terms can be forced to remain in the model using the optional scope argument; see the online help for details. Finally, step returns an object of the same type as its input argument, containing the model chosen by the stepwise search. By default, step displays a table showing the results of dropping each eligible term at each step. This behavior can be disabled by passing the argument trace=F to step. For the auto.lreg object, the results of a stepwise search are shown below; most of the intermediate output has been eliminated.

```
> auto.lreg.step <- step(auto.lreg)
Start:  AIC= 80.103
 Satis ~ Price + Miles.per.gallon + Headroom + Rear.Seat + Trunk +
     Weight + Length + Turning.Circle + Displacement +
     Gear.Ratio + Cargrp


Single term deletions

Model:
Satis ~ Price + Miles.per.gallon + Headroom + Rear.Seat + Trunk +
     Weight + Length + Turning.Circle + Displacement +
     Gear.Ratio + Cargrp


scale:  1
```

	Df	Sum of Sq	RSS	Cp
\<none\>			65.58432	91.58432
Price	1	0.010712	65.59503	89.59503
Miles.per.gallon	1	0.523226	66.10755	90.10755
Headroom	1	0.409295	65.99362	89.99362
Rear.Seat	1	0.525284	66.10960	90.10960
Trunk	1	0.437512	66.02183	90.02183
Weight	1	1.060626	66.64495	90.64495
Length	1	3.520727	69.10505	93.10505
Turning.Circle	1	7.982677	73.56700	97.56700
Displacement	1	0.125000	65.70932	89.70932
Gear.Ratio	1	1.560064	67.14438	91.14438
Cargrp	2	7.238924	72.82324	94.82324

. . .

A table such as the one above shows the sum of squares attributable to the term in question and the effects of removing that term on the residual sum of squares and the C_p statistic. An analysis of deviance of the final selected model shows that most of the nonsignificant terms have been removed:

```
> anova(auto.lreg.step,test="Chisq")
Analysis of Deviance Table

Binomial model

Response: Satis

Terms added sequentially (first to last)
                Df  Deviance Resid. Df Resid. Dev   Pr(Chi)
          NULL                      68   16.81159
        Weight  1 -66.13437         67   82.94596 0.0000000
        Length  1   1.00647         66   81.93950 0.3157504
 Turning.Circle  1  14.63896         65   67.30053 0.0001302
        Cargrp  2   8.61274         63   58.68779 0.0134824
```

For models involving interactions, not all terms are eligible to be dropped from the model at each step. Specifically, no lower-order interactions that involve effects contained in those interactions can be removed until the higher-order interactions themselves are removed. The function **drop.scope** can be used to find out what terms are eligible to be dropped at any point in the model-selection process. For example, the **mtable.11** object in Section 10.7.3 was produced using the following formula:

```
> formula(mtable.11)
Counts ~ Age * Card * Income
```

At the start of the modeling process, you can see the terms that are eligible for removal by typing the following statement:

```
> drop.scope(mtable.11)
[1] "Age:Card:Income"
```

Only the three-way interaction can be dropped at this point. Suppose the three-way interaction was removed as part of the modeling process. We can use the **update** function to modify the formula, and then use **drop.scope** to see which terms are now eligible to be dropped:

```
> new.formula <- update(formula(mtable.11), . ~ . - Age:Card:Income)
> drop.scope(new.formula)
[1] "Age:Card"    "Age:Income"   "Card:Income"
```

Once the three-way interaction is removed, the two-way interactions become eligible for removal. The effects of a stepwise selection on the **mtable.11** object are displayed below, suppressing the intermediate printed object with the **trace = F** argument.

```
> mtable.11.step <- step(mtable.11,trace=F)
> anova(mtable.11.step,test="Chisq")
Analysis of Deviance Table

Poisson model

Response: Counts

Terms added sequentially (first to last)
             Df Deviance Resid. Df Resid. Dev      Pr(Chi)
     NULL                    83    5585.560
      Age   5 5119.572       78     465.988 0.000000e+00
     Card   1  142.161       77     323.827 0.000000e+00
   Income   6  150.061       71     173.766 0.000000e+00
Age:Income 30  114.105       41      59.661 1.000000e-11
Card:Income 6   28.763       35      30.898 6.745569e-05
```

The three-way interaction and the `Age:Card` interaction were removed in the selection process.

Exercises

1. One regression diagnostic that measures each observation's influence on individual beta values is known as dfbetas. For the ith case, DFBETAS$_{ij}$ represents the influence of that case on the jth parameter estimate and is calculated as

$$\text{DFBETAS}_{ij} = \frac{\hat{\beta}_j - \hat{\beta}_{j(i)}}{s_{(i)}\sqrt{(X'X)_{jj}^{-1}}}$$

where the symbols have the same meanings as those in Table 11.1. Write an S function to calculate dfbetas.

2. When producing a `coplot`, it is often helpful to draw lines in each of the dependency panels, to make the relationship between the plotting variables easier to see. Using the data from Section 11.1 that produced Figure 11.4 (or data of your own choosing), draw plots that use the `panel` argument of `coplot` to
   a. draw a straight line in each panel, representing the linear regression between the variables
   b. draw a curve through the data using a polynomial regression
   c. draw a spline through the data using `smooth.spline`

Which type of line gives the best impression of the relationship between the variables in the dependency panels?

3. When a statistical model with two independent variables is studied, it is often useful to produce a contour plot of fitted values from the model, sometimes known as a response surface. Fit a **gam** model to the stack loss data introduced in Section 10.8.1, predicting **Loss** as a function of spline transformations of **AirFlow** and **WaterTemp**. Use the **expand.grid** function to create a grid of **AirFlow** and **WaterTemp** values that span the range of those two variables. Then use **predict** to obtain fitted values at the points on the grid, and **contour** to plot the resulting response surface.

Index

!=, 45
*, 43, 218
+, 43, 218
−, 43, 218
/, 43, 218
:, 18, 218
<−, 4, 127
<, 45
<<−, 122
<=, 45
>, 45
==, 45
?, 2
%*%, 43
%%, 45
%/%, 45
%in%, 218
^, 43, 218
~, 217
−, 5
abbreviate, 70–71
absolute value, 51
access, 181
    direct, 38
adjacent values
    finding means for, 204
AIC, 273
alignment, 176
analysis of variance, 228
ANOVA, 228
anova, 271–272
ANOVA table, 230
any, 153
aov, 215
apply, 147
arc, 207
argument, 51
arguments
    functions, 114
    matching, 114
    named, 124

names of, 125
    variable number of, 123
arguments to modeling functions
    control, 222
    data, 220
    formula, 219
    na.action, 222
    subset, 221
    weights, 221
array, 25
as.matrix, 95
ascending order, 56
ASCII, 55, 172
assign, 39
assignment functions, 127
assignment, 4
assignments
    forcing to .Data, 122
    in functions, 40
    retaining mode, 90
    subset of values, 92
attach, 36
attr, 34
attribute
    user-defined, 35
attributes, 15, 34
.Audit, 173
auditing, 173
    eliminating, 173
axes
    custom, 203
    logarithmic, 198
axis, 203
axis labels, 190
B-spline smoothing, 239
backslash, 17
barplot, 202
barplots
    side-by-side, 202
batch file, 9
beta distribution, 64

binom.test, 251
binomial, 251
binomial distribution, 64
blanks
    embedded, 16
boxplot, 69
bs, 239
C, 113, 130
c, 15
call_S, 136
cancor, 254
canonical correlation, 254
cat, 70, 174
categorical variables, 66, 153, 216
Cauchy distribution, 64
cbind, 106
ceiling, 52
cells, 69
character data, 130
character strings
    abbreviating, 71
    breaking up, 71
character values
    breaking up, 72
    length, 70
    name, 39
characters, 69
chi-square, 251
chi-squared distribution, 64
chisq.test, 251
chol, 53
Cholesky decomposition, 53
class, 166
    removing, 167
classification variables, 216
clorder, 255
cluster analysis, 255
cmdscale, 254
co.intervals, 261
codes, 66
coef, 224
col, 100, 117
column labels, 27
column numbers, 100
command file, 9
command re-editing, 174
commands
    editing, 173
    recalling, 173
    reusing, 174
comments, 10
comparison operators, 45
complex conjugate, 52
complex data, 130
complex numbers, 32, 51
computations
    conditional, 156
concatenation, 105

conditional computations, 156
    function, 156
conditional evaluation, 151
conditioning plot, 259
conflicts, 119
conflicts, 119
contingency table, 68, 151
continuous variables, 216
contour plot, 77, 277
contour, 277
contrasts, 264
    orthogonal polynomial, 264
control
    argument to modeling functions, 222
coordinates
    user, 189
coplot, 259, 276
cor, 61
cor.test, 251
correlated random samples, 53
correlation, 60, 252
    canonical, 254
    squared multiple, 256
    testing, 251
cosine, 51
count.fields, 31
counting, 62, 69
covariance, 60
Cp, 256
crossprod, 54
cut, 66
cutree, 255
.Data, 21
_Data, 21
data
    access, 36
    fixed fields, 30, 40
    insertion, 57
    mixed modes, 28
    organization, 24
    reading, 30
    reading from files, 17
    reading from tables, 237
    storing different modes, 24
data
    argument to modeling functions, 220
data frames, 28, 152
    attaching to search list, 37
    eliminating columns, 109
    numeric columns, 152
    parameterized, 37, 250
    reading data into, 30
    single row or column, 103
data.dump, 172
data.frame, 29
data.restore, 172
database
    updating, 38
    working, 22

decimal places
    controlling, 175
default, 167
default values
    functions, 114, 128
degrees of freedom, 229
deparse, 125
dependent variable, 215
descending order, 56
design matrix, 224, 265
detach, 36
determinant, 54
dev.ask, 211
dev.copy, 210
dev.cur, 210
dev.list, 210
dev.next, 211
dev.off, 210
dev.prev, 211
dev.print, 210
dev.set, 210
device, 75
diag, 53, 111
diagonal, 53
diagrams, 207
diff, 206
digits, 176
dim, 26
dimnames, 27, 95
dimnames, 27
direct product, 55
directory
    data, 21
    home, 21
discr, 254
discriminant analysis, 254
dist, 255
distance matrix, 255
distributions
    discrete, 65
    probability, 63
division
    integer, 45
do.call, 164
dos, 179
dot product, 54, 126
double data, 130
double precision, 130
dput, 172
drop argument to subscripts, 95
drop.scope, 275
dummy matrix, 265
dump, 172
duplicated, 59
dyn.load, 135
dyn.load2, 143
dynamic loading, 130
    libraries, 142
    multiple files, 142

editing commands, 174
editor, 116
ega, 75
eigen, 53
eigenvalues, 53
eigenvectors, 54
ending an S session, 5
environmental variable
    S_PRINTGRAPH_METHOD, 207
environmental variables in UNIX, 180
equality
    testing for multiple, 153
euclidean distance, 255
evaluating text, 165
exact tests, 251
excluding variables, 220
executing commands, 165
exists, 39
expand.grid, 277
expected values, 151
exponential, 51
expression, 165
expressions
    evaluating, 165
    in statistical models, 218
F test, 253
factor, 68
    converting to character, 68
    missing values, 68
factor, 66, 229
file
    counting fields, 31
    external, 171
    input, 17
    MS-DOS, 181
    reading part of, 40
    removing, 123
    temporary, 123
filenames
    MS-DOS, 18
.First, 11
Fisher's exact test, 251
fisher.test, 251
fitted, 224
fitted values, 268
fix, 115, 117, 128
fixed-format data, 40
float data, 130
floor, 52
for loops, 124
format, 175
format, 176
formula
    storing, 219
formula, 224, 271
formula
    argument to modeling functions, 219

FORTRAN, 113, 130
frequencies of values, 69
friedman.test, 251
functions, 49, 113
    arguments, 49
    calling indirectly, 164
    checking arguments, 114
    creating, 117
    default values, 114, 128
    editing, 115
    errors in, 116
    generic, 166
    link, 231
    list of arguments, 164
    return value, 114
    return values, 119
    writing to a file, 172
gam, 215
gamma distribution, 64
gamma function, 51
gen.levels, 238
generalized inverse, 73
generic function, 166
geometric distribution, 64
get, 38
glm, 215
graphics, 75
    displaying on screen, 75
    empty, 76
    flushing, 76
    multiple devices, 210
    multiple figures, 186
    printing, 76
graphics parameters, 183
    adj, 83
    cex, 83, 186
    cxy, 207
    fin, 186
    mar, 186, 202
    mfcol, 77
    mfg, 187, 212
    mfrow, 77
    mgp, 184
    oma, 188, 191
    omd, 211
    omi, 188, 191
    pch, 198
    pin, 186
    srt, 83, 206
    table of, 185–186
    temporary changes, 184
    usr, 207, 189, 195, 202
    xaxt, 201
    xpd, 202
graphs
    adding to, 189
    clearing, 189
    multiple lines, 191

grep, 70, 72, 229
grid
    creating, 277
group labels, 160
grouping, 153
groups
    creating, 66
hclust, 255
header, 30
help
    online, 2
help, 2
hercules, 75
hierarchical clustering, 255
hist, 202
histogram, 202
history
    command, 173
hpgl, 76
hyperbolic functions, 51
identify, 83
identity matrix, 53, 111
if statement, 156
if-else statement, 156
ifelse, 151
Ignoring records in input file, 40
inherits, 167
inner product, 54
input
    multiple lines, 41
integer divide, 45
integration
    numerical, 136
interaction, 217
interaction plot, 200
intercept, 217
    suppressing, 266
interp, 213
interpolate, 213
interpreting text, 165
interrupts, 10
inverse, 55
invisible, 115
is.na, 46
iterative methods, 162
Kronecker product, 55
Kruskal-Wallis test, 253
kruskal.test, 251
kurtosis, 62
labclust, 255
labeling vector elements, 17
labels
    matrix, 27
landscape, 209–210
.Last, 11
latent root, 53
lazy evaluation, 129

ld (UNIX command), 142
leading dimension, 140
leaps, 256
legend, 194
legend, 194
length, 15, 34
    character value, 70
levels, 67
    generating, 238
levels, 66
library, 33
linear algebra, 52
linear models, 224
link function, 231
list, 24, 105
lists
    breaking down, 106
    combining, 105
    converting to vector, 106
    creating from data, 41
    names of elements, 40, 101
lm, 215
lo, 239
location of matrix elements, 117
location of vector elements, 93
locator, 84, 213
loess, 215, 271
log gamma function, 51
log, 51
logarithm, 51
logarithmic axes, 198
logical subscript, 99
logistic distribution, 64
logistic regression, 232
    data from tables, 234
lognormal distribution, 64
loops, 147
Mann-Whitney test, 253
Mantel-Haenszel chi-square test, 252
mantelhaen.test, 251
mapping, 147
masked, 119
match, 58, 119, 158, 229
matching
    partial, 114
matlines, 192
matplot, 191
matpoints, 192
matrices, 25
matrix labels, 27
matrix
    accessing elements, 94
    adding rows or columns, 106
    and dynamic loading, 139
    arithmetic operators, 44
    concatenating, 106
    creating from table, 99
    deleting rows or columns, 161

design, 224
diagonal, 53
distance, 255
filling with zeroes, 25
functions, 52
identity, 53, 111
mapping functions to, 147
model, 224, 265
multiplication of, 44
plotting columns, 191
processing by groups, 154–155
reshaping, 110
singular, 73
size, 26
storage of, 26
subsetting, 94
symmetric, 255
transpose, 110
variance-covariance, 61
zeroing elements, 100
matrix, 25, 111
max, 60, 62
maximum, 60
McNemar chi-squared test, 252
mcnemar.test, 252
mean
    columns of matrix, 150
memory allocation, 140
merging data, 58
messages, 175
mfcol, 77
mfrow, 77
min, 60, 62
minimization, 256
minimum, 60
missing, 114
missing values, 139, 149
    comparisons, 46
    eliminating, 93
    factors, 68
    logical expressions, 46
    modeling functions, 222
    patterns, 164
    recoding, 151
    subscript, 90
    subsetting expressions, 97
    testing for, 46
mode, 15, 34, 39
    in search, 23
model
    statistical, 217
    updating, 227
model matrix, 224
model selection
    stepwise, 273
model.matrix, 265
models
    expressions in, 218

modes
    mixed, 28
modulus, 45, 52
motif, 75
moving to a new computer, 172
MS-DOS, 179
ms, 215
mtext, 202
multiple graphics devices, 210
multiple plots, 77
multiple values, 59
multivariate normal, 53
na.action
    argument to modeling functions, 222
na.rm, 60
names
    abbreviating, 92
    access by, 39
    finding duplicate, 119
    function, 4
    object, 4
    of list elements, 160
    of object, 195
    suppressing printing, 177
    variable, 4
names, 17
nchar, 70
ncol, 26
negative binomial distribution, 64
nested effects, 218
newline, 175
Newton's method, 162
nls, 215
nonatomic, 24
normal distribution, 64
normality
    graphical test, 259
nrow, 26
ns, 239
NULL, 161
numbers
    format, 175
    random, 19
object, 15
object-oriented programming, 165, 215
objects
    combining, 105
    existence of, 39
    information about, 23
    storing, 21
objects, 22
objects.summary, 23
online help, 2
openlook, 75
operating system, 179
operators
    arithmetic, 43
    as functions, 125

logical, 45
    precedence, 47
    special, 47
options, 12
or
    exclusive, 46
order, 56
orthogonalization, 54
outer, 54, 151, 206
outer product, 54, 151
outliers, 267
output
    printing, 76
paired test, 252
parameterized data frame, 37, 250
parameters, 249
    graphic, 183
    resetting, 184
parentheses, 49
parse, 165
partial matches, 114
partial matching, 59
partial sums of squares, 263
paste, 69–70, 121, 179
percentile, 60
permutation, 62
permutation vector, 56
persp, 204
pframe, 250
phase angle, 52
pie chart, 77
plclust, 255
plot
    contour, 77
plots
    3-D, 204
    adding text, 190
    annotating, 83
    conditioning, 259
    interacting with, 83
    interaction, 200
    locating points, 84
    multiple, 77
    perspective, 204
pmatch, 59, 114
pointwise, 224
Poisson distribution, 64
polynomial regression, 269
PostScript, 207
    resolution, 209–210
postscript, 76
prcomp, 254
precedence
    operator, 47
predicted values, 224, 268
pretty, 66
principal components, 254
print, 10

`printgraph`, 207–208
printing
    suppressing in function, 115
    to file, 175
printing graphics output, 76
printing without quotes, 17
printing, 174
probability distributions, 63
probability, 63
prompt, 4
`prop.test`, 251
proportions, 251
q, 5
`qqnorm`, 259
`qr`, 54
`quantile`, 61, 67
quantiles, 64
quit, 5
random number, 19
random number generation, 65
random samples, 61
    correlated, 53
`.Random.seed`, 65
randomized block design, 253
`range`, 62
`rank`, 57
rank sum test, 251, 253
rank test, 251
`rbind`, 106, 161
`read.table`, 30
reading data from files, 17
recalling previous commands, 173
recoding, 158
records
    skipping, 40
recursive objects, 106, 24
regression
    all subsets, 256
    leaps and bounds, 256
    linear, 224
    logistic, 232
    polynomial, 269
regression diagnostics, 267
regressors, 216
`rep`, 19
repeated values, 19
`replace`, 58
replacement functions, 127
reshaping matrices, 110
residuals, 224
    generalized linear models, 232
`residuals`, 224
response surface, 277
`restore`, 172
`rev`, 19
reversing elements of a vector, 19
root, 53
roots, 162

rounding, 45, 52
`row`, 100, 117
row labels, 27
row numbers, 100
`rug`, 241
`s`, 239
`sample`, 61
samples
    random, 61
saving detached objects, 37
saving graphics parameters, 184
scalars
    storing in data frame, 37
`scale`, 55
scan, 40
    reading multiple lines, 41
    skip argument, 40
`scan`, 16, 123, 41, 111
scanning output of UNIX commands, 123
`search`, 22
search list, 125
    removing from, 36
search path, 22
separators, 17
`seq`, 18, 93
sequence operator, 18
sequential sums of squares, 263
`set.seed`, 65
`sign`, 51
signed rank test, 251
significant digits, 52
signum function, 51
Simpson's rule, 136
simulations, 160
sine, 51
single, 130
singular value decomposition, 54
singular, 73
`sink`, 171
size of a matrix, 26
skewness, 62
skipping records, 40
`smooth.spline`, 192
smoothing, 192, 239
`solve`, 55
`sort`, 60
`source`, 10
spline, 192, 239
    B-, 239
    natural cubic, 239
splines, 239
`split`, 66, 181, 255
square root, 51
squared multiple correlation, 256
standardization, 55
standardizing, 55
statistical models
    comparing, 270
    updating, 270

stem, 20
stem and leaf diagram, 20
stepwise model selection, 273
stopping S, 10
strings
    collapsing a vector, 70
    concatenation, 70
structure, 35
Student's t distribution, 64
Student's t test, 252
subscripts, 89
subset
    argument to modeling functions, 221
subset of values
    changing, 92
subsetting, 153
subsetting a matrix, 94
substitute, 125
substring, 70–71, 127
subtree, 255
sum
    cumulative, 60
sum, 62
sums of squares
    partial, 263
    sequential, 263
superdiagonal, 98
svd, 54
switch, 158
symmetry
    matrix, 112
synchronize, 38
system commands, 179
system of equations, 55
S_PRINTGRAPH_METHOD, 207
t, 110
t test, 251
t test
    unequal variances, 253
t.test, 251
table
    ANOVA, 230
table, 68
tables
    printing, 177
tail probability, 63
tangent, 51
tapply, 255
termination
    emergency, 10
test
    one-sided, 251
    paired, 252
text
    adding to plots, 190
time series, 32
timing commands, 191

title, 190
transport, 172
transpose, 110
tree, 215
tree-based models
    prediction, 269
trigonometric functions, 51
trunc, 52
truncating, 45, 52
truncation, 52
unary minus, 45
unbalanced ANOVA, 263
unclass, 167
uniform distribution, 64
unique, 59
unix, 179
UNIX commands
    echo, 180
    scanning output, 123
    test, 181
    wc, 179
UNIX environmental variables, 180
unix.time, 191
unlist, 106
update, 224, 227, 275
UseMethod, 166
user coordinates, 189
values
    repeated, 19, 59
var, 61
var.test, 251
variables
    categorical, 66, 216
    classification, 216
    continuous, 216
    dependent, 215
    independent, 215
    local, 121
variance
    testing equality, 253
variances
    comparing, 251
vector
    arithmetic operators, 43
    as matrix, 95
    changing a subset, 92
    column, 96
    labeling elements, 17
    permutation, 56
    row, 96
vga, 75
vi, 116
Weibull distribution, 64
weights
    argument to modeling functions, 221
Welch modified t test, 253
wilcox.test, 251
Wilcoxon tests, 251

`win.graph`, 75
`win3`, 179
Windows, 179
working database, 22
    changing, 37
`write`, 171
`x11`, 75
z-score, 150
zero
    subscript, 90
zeroes, 162